River Odyssey

River Odyssey

A STORY OF THE COLORADO PLATEAU

by Gerald N. Callahan

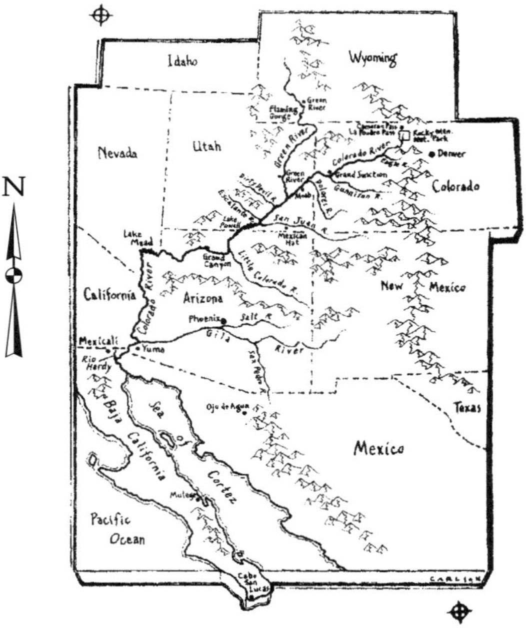

University Press of Colorado

ISBN: 0-87081-469-9

Published by the University Press of Colorado
P.O. Box 849
Niwot, Colorado 80544

Printed in the United States of America.

The University Press of Colorado is a cooperative publishing enterprise supported, in part, by Adams State College, Colorado State University, Fort Lewis College, Mesa State College, Metropolitan State College of Denver, University of Colorado, University of Northern Colorado, University of Southern Colorado, and Western State College of Colorado.

The paper used in this publication meets the minimum requirements of the American National Standard for Information Sciences—Permanence of Paper for Printed Library Materials. ANSI Z39.48-1984

Library of Congress Cataloging-in-Publication Data

Callahan, Gerald N., 1946–
River odyssey : a story of the Colorado Plateau / Gerald N. Callahan.
p. cm.
ISBN 0-87081-469-9 (hardcover)
1. Colorado Plateau—Description and travel. 2. Colorado River (Colo.-Mexico)—Description and travel. 3. Callahan, Gerald N., 1946– . 4. Hiking—Colorado Plateau. I. Title.
F788.C16 1998
917.91'3—dc21 97-44141
CIP

10 9 8 7 6 5 4 3 2 1

For my wife, Gina, who in many ways is the river
and who made all of this possible.
For my son, Patrick, and my daughter, Jennifer,
who have shown me things I would have missed.
And for my mother, June, and my father, Wayne,
who gave me fingers that I might write.

Irene
Thank You For
your Support, Fellow
Writer

[illegible]

CONTENTS

Introduction	1
Prologue: Uncle Henry's Craziness	5
Watermarks	9
She Said Forests	19
In the Beginning	20
The Snag at Potato Bottom	26
Desert Sonata	27
Near Gypsum Gap	34
Grand View Point	35
The Way Things Are	44
Almost Like Dancing	45
The Dog at Christ's Feet	56
In the End Death Looks a Lot Like Me	58
Natural Selection	64
Because I Am Mad About Women	66
The Whiteness of Birds	74
Cheese and Raisins	76
Dumpster Diving	86
Ravens	87
Raven's Dream	97
My Canyon	98
Vows	110
The Dead Man Near Monument Valley	111
The Thing Inside Your Head	115
Death Wish	117
Unemployed in Greenland	130
Aravaipa Canyon	132
Autobiography	140
Yellow-Crowned Night Heron	142
A Winter's Solstice	152
Snowy Egrets: The Ending	154
Watermarks	173

LIST OF PHOTOS

The photo of the author in the San Rafael Swell was taken by Dana Leach. All other photos were taken by the author.

PAGE	PHOTO
10	Ray's Tavern, Green River, Utah
15	The San Rafael Knob atop the San Rafael Swell
20	La Poudre Pass, Colorado—the Beginning
27	The author's brother, Michael, and his favorite form of transportation, near Kanab, Utah
33	The Grand Valley from the mesas near Bull Canyon
38	Dead Horse Point and the potash evaporation ponds, Dead Horse Point State Park, Utah
41	Grand View Point, near Moab, Utah
43	The author's son, Patrick Callahan, walking on thin ice near Moab, Utah
49	The Professor Valley, on the road from Cisco to Moab, Utah
50	Canyon reflections—Entrada sandstone near Moab, Utah
51	Jug Handle Arch, near Moab, Utah
60	Deep inside the Colorado Plateau
62	Author approaching end of line near Hanksville, Utah
67	White Canyon, near Hite, Utah
72	Fry Canyon, near Fry Canyon, Utah
77	The author's daughter, Jennifer Callahan, near Blanding, Utah
78	The author in the San Rafael Swell
90	Glen Canyon Dam, Page, Arizona
93	The Colorado River where it enters Lake Powell, near Hite, Utah
98	Pat and Steve Larkin near a cloud-filled Grand Canyon, Arizona
103	The author's father, Wayne Callahan, at The Neck, near Moab, Utah
105	Mr. Peabody's coal chute, near Kayenta, Arizona
108	Reflections in a stone tank inside my canyon
112	Monument Valley, Arizona
118	The last of Glen Canyon, near Page, Arizona
127	The author exploring a tributary of Glen Canyon
133	Aravaipa Canyon, near Safford, Arizona
137	Another view of Aravaipa Canyon
142	Les Walker, the card cheat (a clean-shaven incarnation), on the road to Mulege
148	The Rio Mulege near Mulege, Baja California

157 The "Gunderson" River, near Grand Junction, Colorado
159 The Dolores River, near Naturita, Colorado
160 The Disappointment Valley, near nowhere, Colorado
162 The San Juan River at Mexican Hat, Utah
163 Indian "rock art," near Mexican Hat, Utah
171 The bus from Ledon to Luis B. Sanchez and beyond crosses the Colorado River near its end

ACKNOWLEDGMENTS

Citations

Emily Dickinson: From poem #712 in Thomas H. Johnson, ed., *The Complete Poems of Emily Dickinson,* with permission of Little, Brown and Company.

Homer: From *The Iliad.* Translated by Robert Fitzgerald. Published by Alfred A. Knopf, Inc., 1992.

Langston Hughes: From "The Negro Looks at Rivers" in *Collected Poems,* by Langston Hughes. Copyright 1994 by the estate of Langston Hughes. Reprinted by permission of Alfred A. Knopf, Inc.

Galway Kinnell: From "The Shoes of Wandering" in *The Book of Nightmares.* Copyright 1971 by Galway Kinnell. Reprinted by permission of Houghton Mifflin Co. All rights reserved.

Norman Maclean: From *A River Runs Through It and Other Stories,* by Norman Maclean. Published by The University of Chicago Press, 1976.

Herman Melville: From *Moby Dick,* by Herman Melville. Easton Press edition, 1977.

Pablo Neruda: From "Amor America" in *Canto General,* by Pablo Neruda. Translated and edited by Jack Schmitt. Copyright 1991 by Fundacion Pablo Neruda, Regents of the University of California. Reprinted by permission of University of California Press.

Naomi Shihab Nye: From "Negotiations with a Volcano" in *Words Under the Words: Selected Poems,* by Naomi Shihab Nye. Published by Far Corner Books, Portland, Oregon. Copyright © 1995 by Naomi Shihab Nye. Reprinted by permission.

Rainer Maria Rilke: From "I Live My Life" in *Selected Poems,* by Rainer Maria Rilke. Copyright 1981 by HarperCollins Publishers, New York. Reprinted by permission of HarperCollins Publishers.

Alberto Alvaro Rios: From "The Waltz of the Fat Man" in *Pig Cookies,* by Alberto Alvaro Rios. Published by Chronicle Books, 1995.

Patiann Rogers: From "The Creation of the Inaudible" in *Firekeeper,* by Patiann Rogers (Milkweed Editions, 1994). Copyright 1994 by Patiann Rogers. Reprinted with permission from Milkweed Editions.

Carl Sandburg: From "Wilderness" in *Cornhuskers,* by Carl Sandburg, copyright 1918 by Holt Rinehart and Winston, Inc. and renewed 1946 by Carl Sandburg. Reprinted by permission of Harcourt, Brace and Company.

Leslie Marmon Silko: From *Ceremony,* by Leslie Marmon Silko. Published by Viking Penguin, 1977.

Publications

"The Snag at Potato Bottom" originally appeared in *Phase and Cycle.*

"Natural Selection" and "The Thing Inside Your Head" were first published by *Midland Review.*

"The Whiteness of Birds" and "Unemployed in Greenland" were originally published by *Southern Poetry Review.*

"The Way Things Are" originally appeared as "Fairy Tales" in *Rhino.*

"A Winter's Solstice" was first published in *Midwest Quarterly.*

"Raven's Dream" and "Watermarks" first appeared in *The MacGuffin.*

"Almost Like Dancing" first appeared in *Cream City Review.*

"Yellow-Crowned Night Heron" was first published by the *Pacific Review.*

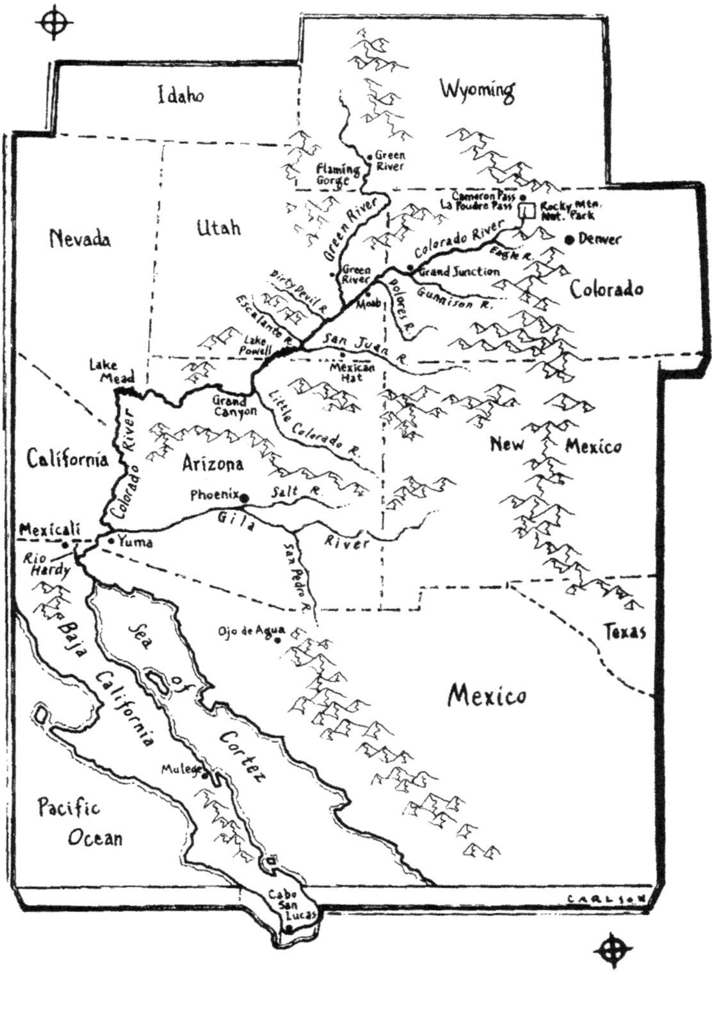
Idaho
Wyoming
Green River
Flaming Gorge
Cameron Pass
La Poudre Pass
Rocky Mtn. Nat. Park
Denver
Nevada
Utah
Green River
Colorado River
Eagle R.
Grand Junction
Colorado
Dirty Devil R.
Green River
Moab
Dolores R.
Gunnison R.
Escalante R.
Lake Powell
San Juan R.
Lake Mead
Mexican Hat
Grand Canyon
Little Colorado R.
New Mexico
California
Colorado River
Arizona
Phoenix
Salt R.
Gila
River
Mexicali
Yuma
Rio Hardy
San Pedro R.
Texas
Sea of Cortez
Ojo de Agua
Baja California
Mexico
Mulege
Pacific Ocean
Cabo San Lucas
CARLSON

River Odyssey

Introduction

It's 5:30 on a Saturday morning, hard into the Colorado winter. An hour ago my feet felt like I imagine Maurice Herzog's feet must have felt up on Annapurna—just before most of his toes had to be removed. My feet don't feel like that now—right now they don't feel like much of anything. My fingers feel, though, like handfuls of frozen bees. Mostly the sky is dark, a wombful of stars. But east of here the cauldron's lid has been lifted, and through the crack I can see the final flame. It will be months, though, before the warmth of that fire reaches me.

I'm lying flat on my back among the furrows of an old cornfield, my head propped on an ammo can. While I watch the blaze at the edge of the earth my gloved left hand clutches a cup of steaming coffee and my bare right hand strokes the Parkerized barrel of the twelve-gauge next to me where it catches the first dull gleams of morning. But mostly I'm listening—listening for the lunatic laugh of geese.

My wife, Gina, has wrapped herself in a camouflage quilt next to me and has moved off into her own thoughts, or sensibly gone back to sleep. Out here, at the westernmost edge of the Great Plains, sunrises are dramatic, especially when the fields are snow-covered. Then, each day begins explosively. But snow or not, the great peaks of the Rockies' Front Range—Longs, Meeker, Storm, Dragon's Egg—are the first earth to catch each day's new heat. That light begins nearly crimson, puddled among the ice fields and wind slab that drape the rigid tips of the winter Rockies. Then the day moves, gathering speed, down the 9,000-foot faces—red, to gold, to ice white—and finally spills out across the plains, the corn stubble, the rutted roads, the iced ponds—the color now of earth.

Up there, among the frozen spires and crumbling stone, as each day begins the waters of this continent assemble. Then, carefully as Moses, they divide themselves into rivers. Rivers that will this day and on all others to come, change the lives of every living thing, and most dead things, from here to the Gulf of Mexico, from here to the Sea of Cortez.

 The South Platte, the North Platte, the Arkansas, the Rio Grande, the San Juan, the Colorado, the Gunnison, the Dolores, the Yampa, the Eagle, the White, the Blue, gathering themselves up there in this morning's light.

Gina and I came to Colorado twelve years ago along with two children, a miniature schnauzer with the social skills of a scorpion, and a bookful of expectations. Patrick has his own place in town now, and Jennifer lives in Minneapolis. Our schnauzer, Nellie, died in my arms two years ago, and mostly we've decided the book—the one full of expectations—has to be rewritten. Or maybe the land or the rivers or the Gambler decided the book needs to be rewritten. It makes no difference—things have changed, that's all.

That's where *River Odyssey* came from. It began five or six years ago among the peaks where the day's first fire is dancing now. I was walking a trail once called the La Poudre Pass Trail, now called the Colorado River Trail. That trail follows the first seven or so miles of a growing stream that becomes the Colorado River—though there remains some dispute over just where that happens.

A couple of years before, a friend and I had spent a winter week in San Felipe, near the northern end of the Sea of Cortez. As I stood there, on the La Poudre Pass Trail, I remembered the tidal flats at San Felipe spread like San Pasqual's apron across northern Mexico. And I realized that the Colorado was the first river I'd seen from both ends. That pleased me. But as quickly, I realized I knew nearly nothing about all that lay between where I stood and the place where the river died. There was a gap.

My life just then was Swiss-cheesed with gaps—my job, my perspective, my sense of how any of it fit together. And in the decreased oxygen of that 10,000-foot-high trail and the blueberry shade of the Never Summer Mountains, I made a decision. A decision to close some of those gaps. And I chose the river as a place to begin. That seemed right, to the river and me.

Somewhere along the way I started writing the words down. And somewhere else along the way I realized I was doing more than just writing. Then, standing one afternoon in an ice-cold plunge pool deep in the Colorado Plateau, I got the idea of putting what I'd written into a book. A book in defense of all this—humanity, rivers, canyons, fear, depression, aging. But that isn't what the book became. Nothing so powerful as this earth or as inevitable as human craziness needs a defense. The book became something else. Something with tree toads in it and the dark of the sky at night. Something that smelled like desert rain and wet creosote. Something necessary—for me, at least.

River Odyssey is about time I've spent, mostly on the Colorado Plateau, looking for those gaps I mentioned, the spaces in the cheese of the world. The story isn't finished, of course. Hardly any story ever is. But a part of this story *is* finished, the part about how it began and the part about the river—the same river gathering itself now, up there where the sun's light ricochets off panes of water ice and crashes like elephant ivory through frozen cornices—mostly, the part about the river.

I hear them. And now I see them. A wedge of morning-gray geese coming in low over the barbed-wire fences to the north. Each with a wing in sunlight and a wing in darkness—hemi-lucent. Just ahead of the decoys the birds lift, turn, and begin a long circle. Canadas are miraculous in flight—heavy-bodied, full-winged, and a grace born of the north wind itself. Their laughter now is for me.

Gina begins to flag. I work the call briefly, then stop, and quietly slide the safety off on the shotgun. The birds pull their wings into a full deep cup, drop their palmated feet, and fall, silent as stones, into the deceit of our decoys.

Prologue: Uncle Henry's Craziness

My Uncle Henry was crazy. Even as a boy I knew that. Mom had told all of us kids that her brother was crazy. What I didn't know then was that it was syphilis—spirochetes—burrowing into Henry's brain that made him crazy. Mom didn't mention that part. Henry was with the infantry during World War II. But as I recall, he didn't receive his decoration in battle. It was somewhere else along the line—in a hayloft or feather bed or bleach-sheeted hotel room—that Henry was awarded the iron cross of insanity.

By 1956 the spirochete was regularly getting between Henry and what most people would consider proper behavior. That year the army took over Henry's care and shipped him to the VA Hospital in Salt Lake City, Utah. In 1950 my family had moved from Coffeyville, Kansas, to Bountiful, Utah—about twelve miles north of Salt Lake City. So, in spite of the army, Henry ended up near family. In 1956 I was ten. By the time I was thirteen or fourteen, Henry was dead. But in the few years between, nearly every Sunday, Mom and Dad got Henry from the hospital and brought him out to the house for dinner.

Even after considering my mother's warnings, I thought Henry was one of the more interesting adult human beings I'd met—though not when my friends were around, of course. Henry often looked and smelled somewhat unusual, unusual for an adult male human anyway. I was uneasy about how my friends might react to that. Not because of how it would affect Henry. He probably cared not one whit what my friends thought of him. No, my concern was how his appearance and aroma would affect me and my social standing. My concern was what sort of grist Henry, in all his glory, might provide for the stories my friends and I loved to shame one another with. So I once refused even to recognize Henry when a group of us passed him on his way down Second South in downtown Salt Lake. That's a thing I've never forgotten. Never will. I still see him at times working his way on down that Salt Lake street, rumpled, confused.

But when my friends weren't around I thought Henry was damned interesting. And it's fascinating to me, even now, the things Henry chose to do with his gift.

Henry was a tall man, over six feet, and skinny as a whippet. And he always looked like someone had left the hanger in his shirt. Henry was mostly bald, with an unruly fringe that strung out in the breeze like rye in late fall, and his once-blue eyes had moved far back into the hollows of his sockets. He looked a lot like my grandfather in the pictures on my grandmother's dresser, though grandpa was years older when those photographs were taken.

Because of the VA laundry, Henry's clothes were always well ironed, but his shirts were rarely buttoned correctly. His fly was almost never zipped, and the crotch of his khaki pants was often stained yellow with urine from a penis regularly withdrawn a little too late from or returned a little too early to its nest of wiry gray and blond hair. I never saw Henry cleanshaven, but then I never saw Henry with a beard, either.

Henry's shoes were a curiosity to me. I'd never seen footwear like his. Tennis shoes, I guess you'd call them. But no one ever played tennis in shoes like those. Certainly Henry didn't. Henry didn't play anything remotely like tennis—the spirochete had seen to that. Coarse canvas uppers the shoes had, and pecan-colored crepe rubber soles pimpled like peanut brittle. No laces—these were slip-ons, with little bellows of fabric stitched to the tongues. The only thing remotely athletic I could imagine someone doing in shoes like those was tossing horseshoes or tottering across the checkerboard patterns in tiled hallways. Those were old-man shoes. Those were asylum shoes.

I didn't think of Henry as particularly old, though. And I didn't really think of him as crazy, either. But I didn't care much for the shoes.

Looking back, I suspect the syphilis was in Henry's spine as well as his brain, because when he walked, his gait was a little tentative, a little spastic. Henry seemed to be pulled forward by his front leg—rather than propelled ahead by his back leg, as most of us are—a little like a stork.

Henry's speech was almost always animated. And even though I could only follow a little of what he said, Mom and Dad acted as though they understood most of his words. But at times Henry would just sit, off to the side of the patio or at the gray Formica table in the kitchen, and carry on slow, muttering conversations. It seemed at first he was simply talking to himself. But as the weeks and months passed, I noticed Henry often acted as though he were talking with someone else, someone I couldn't see. I wished even then that I could make out more of those conversations, catch some of what Henry was saying.

After dinner, Mom and Dad would gather Henry's things—a sweater, a pack of Pall Malls, a cracking leather toilet bag—and Henry would brighten again. He would come to each of us children, shake our hands, and call us things like Squirt, or Geronimo, or Daisy Mae. Then he'd climb into the car with my parents and disappear for another week or two. As I said, a damned interesting human being—especially in a family laced with traditional Irish Catholicism, and a family where one of the worst things imaginable was to be perceived as incompetent or a fool.

Now, some thirty-five years after Henry's death, I'm again fascinated by my uncle. You could argue, surely, that some of Henry's idiosyncrasies were just a consequence of the spirochete howling inside his head. But you could argue just as convincingly, to my mind anyway, that Henry chose some of those habits. Chose them once he could see that people would no longer judge him the way they judged others, chose them once he perceived the gilt edge of the spirochete's gift.

So today, sitting in my office—where everyone, including me, thinks I should be writing a grant proposal—I'm thinking. If I were given my uncle's gift, if I were offered the freedom of the insane, what would I do with it?

Well, the first couple of things would be easy. My uncle already showed me. I wouldn't bother zipping my fly—seems a waste of time, and I don't believe anyone but my wife much cares what's behind it anymore anyway. I'd stop worrying about those urine stains, too. I think I'd skip the crepe-soled canvas slip-ons, though—old-man shoes. But I'd get skinny like Henry, and maybe bald too—just easier. And I'd do more of those things people already think I'm crazy for doing—writing, hiking alone, drinking too much beer, and, of course, muttering. On Saturdays I'd go for long walks alone, and on Sundays I'd sit out back, smoke cheap cigars, drink vodka gimlets, and listen to Patsy Cline. I wouldn't wear socks.

And the next time I passed Henry on a street anywhere, I'd step away from my friends, walk right up to Henry and say, "Hello Uncle." Then I'd lie to him, tell him the last time was an accident, tell him I wasn't certain it was him that day in Salt Lake City (and he'd know I was lying). When I was finished lying, I'd tell Henry that I was fascinated by him, by his shoes, his walk, all of the hair in his nose and ears, even his open fly. I'd tell him I hoped someday to be a lot like him (and he'd know that was the truth).

 Then maybe I'd give Henry a hug, rub my stubbly cheek against his. Maybe he'd hug me back, forgiven. Then I'd tell Henry I hoped my Mom would invite him to dinner more often. And my group of friends and the passersby wouldn't even mind or think less of us, because they'd know me and Henry were crazy.

. . . but on the grip
of his weapon of moist flint,
the initials of the earth were
written.

No one could
remember them afterward: the wind
forgot them, the language of water
was buried, the keys were lost
or flooded with silence or blood.
—Pablo Neruda

Watermarks

I no longer remember when I first noticed. Perhaps it was on one of those nights I stole away from the world of people and swam in the tepid ink of a night sea. Perhaps it was an afternoon standing above the arch at Land's End watching water turn desert stone into sand. Or maybe it was the day my daughter was born in a spasm of muscle and salt. In truth, it no longer matters.

We are not earth creatures who occasionally require a little water. We are, each of us, water beings—two- or four- or no-legged cruets of seawater trying to make our way in an earthen world.

The proof sits like a yellow scorpion in the warm sand of the Sonoran, or Mojave, or Chihuahuan Deserts. When ambient temperature creeps significantly above body temperature, lack of water causes dementia in men and women within hours, and death within a day. Even loneliness takes longer than that to kill human beings. Every chemical reaction that distinguishes a living thing from a dead thing depends one way or another on water. Without it, we are simply heaps of salt and carbon, phosphorous and sulfur—ash, waiting to be scattered by the first dry breeze. Creatures of water we are, each of us. And me more than most.

 Just now, I'm taking on one more bottleful of that vital liquid through the kindness of a barmaid by the improbable name of Shirley inside a dark little dive in Green River, Utah. Never mind the fermented barley malt, the hops, the fizz, and the 3.2 percent ethyl alcohol. It's the water I'm after. And thanks to the folks at Budweiser, what I'm drinking tastes more or less just like water. In fact, it's likely this beer was brewed in Fort Collins, Colorado, and that means it's probably mostly the same water—a fifty-fifty mix of Colorado and Mississippi River water—that comes from the taps in my home. That kind of warms me, as does the

Ray's Tavern, Green River, Utah

sight of Shirley's denim-clad behind as she bends to lift more beer glasses from their place below the bar this night in Green River.

"It's all going to hell," she says as she straightens. "Shit, they're even building a golf course. Can you imagine that? A golf course in goddamn Green River, Utah." Shirley shakes her head in disgust as she wipes dust from the glasses with a particularly foul-looking bar rag. "And I heard last week about a house with only three bedrooms and no more'n an acre of land around it sold for near twenty-five thousand dollars. You better go see it all now darlin', 'cause it ain't gonna be around much longer!"

I had just come across the Swell after a hike in Devil's Canyon. The sun was playing its low-angle games with the canyons and mesas, spires and buttes and cracks of the San Rafael Swell and the Eastern Reef. And as I drove I–70 into Green River, the desert throbbed with colors like there will only be again on the very last day, the day Gabriel blows his final riff and the sun goes red giant, the day we all get toasted like gummibears on a cheap hot plate. I understood full well what was eating at Shirley Well, maybe not full well—she lives here, her love is greater even than mine. But I felt a reasonable approximation of what was gnawing at her.

"But there's no water around here that isn't already committed to Arizona or California, or Mexico. That's got to slow them down," I offer. Then, as visions of the metastatic flare of Moab congeal somewhere between my parietal and temporal lobes, even I don't believe what I just said. Mostly though, I just want to keep Shirley making words in those odd shapes they acquire inside Utahans' mouths.

"Don't bet on it sweetheart," she said. "Number one, the Californians and God-knows-what-else that're movin' in don't have sense enough to worry about water. And number two, the goddamn developers are already figurin' out ways to use more of the Green. And if that ain't enough, the place is fillin' up with foreigners too."

I doubt there's a bigoted neuron, white blood cell, or adipocyte in Shirley's pert little body, but when you're forced to sit and watch while money-thirsty men sodomize the things you love, even the strong can say things they don't truly mean. I nudge the conversation toward the real enemy.

"I never thought it would happen here, not in my lifetime anyway. Human greed is greater than even I'd imagined, and I have a pretty vivid imagination. But developers reproduce faster than roaches."

"Where there's a buck to be made. . . . Get used to it, love," she advises.

We both shake our heads, then Shirley moves off to grill a couple of cheeseburgers for two cowboys who have settled with their boots and their blue jeans into a booth behind me. Nailed into the imitation mahogany paneling where Shirley just stood is a picture of several people that might include Shirley in a big rubber raft nearly on its front end in the middle of a standing wave the color of coffee—Lava Falls maybe. Everyone appears to be enjoying themselves immensely. Water creatures, all of us.

And as I said, me more than most. Two observations have driven me to this conclusion. First, I need, and I mean *need,* to be close to large bodies of water, often. Right now that's the Colorado and its twisted tributaries. The fact that there are only two other people in this entire bar (a few short blocks from the glorious Green River) on the twentieth of November suggests to me that the water needs of others, if not less than mine, at least are manifest only in fair-weather months. And second, when I was three years old I was scalded like a tomato by boiling water. A stupid accident, but it cost me near half my skin. There's still a thick scar about the size of a big plate of enchiladas and shaped a little like North America—with Nova Scotia nestled among my middle ribs and Florida flung across the top of my left thigh. And I've spent enough self-conscious moments in men's dressing rooms to know most people haven't been touched by water like that.

I think about water a lot—oceans and rivers and ponds and creeks, wretched swamps and bilious bogs, polar ice and bottles of Budweiser. But lately it's been mostly the Colorado River. Since I moved from the West Coast, the river is the only wet thing near big enough to make me forget about the ocean. Forget at times, anyhow.

By the way, I don't want to leave you with the wrong impression here. I am not a native southern Californian. But I did serve a ten-year sentence there after growing up near Salt Lake City, Utah. And although California does have its drawbacks, there is nothing in this whole goddamn world like the Pacific Ocean.

Shirley offers to buy my next beer. I accept, graciously. She draws a cold one, then lights a filter-tip Camel cigarette, pushes a wisp of brown hair from her forehead, and watches, with a hand at her hip, the cowboys from behind curls of tobacco smoke. Thirty-two or three maybe, slick in her sweater top and tight blue jeans. A little too much eye shadow maybe, but maybe not.

My mother was told more than once, in the days after I was burned by water, that I wouldn't survive, that my first dance with feral water would be my last. Those who told her this were wrong. But it did take near six weeks flat in bed while they pumped salt water into me through steel needles stuck in veins near my clavicles before anyone else came to believe what I'd known pretty much from the outset. Water wasn't done with me.

The land here is like that, watermarked. Today I walked the upper end of Devil's Canyon, up on top of the Swell. Usually a less than grand idea in winter, but today the cold wasn't so bad. The road dumps you off smack in the middle of the canyon, and you can go up or down. Last December a friend and I walked some of the lower portion of this canyon. So today I thought I'd take a look at the upper end and the San Rafael Knob—I'm kind of into knobs.

The walk up-canyon was easy, except for one spot where I ran into a fifty- to seventy-five-foot-high dry falls that was unscalable. I had to walk back down-canyon for a couple hundred yards before I could climb out and skirt the falls. I stuck a juniper branch into the sand where I reentered the canyon to help direct me on the way back. Other than that, there were just a few abrupt little cliffs in the Kayenta, then Navajo sandstones, where I tossed my pack up and then climbed the yard or two to the next level. The expanses of stone, the lack of wildlife this time of year, and the silent blue-and-white cameo of clouded sky overhead left lots of time for thought.

Piñon trees seem a stoic lot. I imagine them without purpose, except perhaps to witness days such as these, a stellar flare somewhere beyond the Pleiades, the occasional death of a desert hare, the reluctant blue of the sky at evening, and the swarms of intricate dreams left here for those who will never claim them. Stoic surely, but not likely bored.

The canyon, as always, seemed longer than the map indicated.

It was twelve or thirteen months before the burned skin could be left undressed. 'Til then, twice a week a registered nurse came to our house to change the crusted gauze and wonder at the power of flesh to heal. Each time she came I cried out loud and hid behind my mother, then screamed while she peeled away the old gauze and taped on new. It wasn't because of the pain, though, and the nurse knew that. So one day she showed my mother how to change the dressings. I mostly quit bawling and carrying on then, and quit altogether when my mother lost the skin-eating adhesive tape and used one of her girdles instead to hold the dressings in place. The doctor admired that, admired my mother's ingenuity. And slowly, something that looked a little like skin began to replace large chunks of red and black ruin.

About two hours into Devil's Canyon the Navajo sandstone opens wide. To the southeast you can see the San Rafael Knob. If you continue up-canyon here you can drop over into Eagle Canyon, cross under I–70, and move into the eastern half of the Swell toward Sids Mountain. But today I was interested in the Knob and the view from up there. So I moved out cross-country.

Snow still stood in the shadows beneath the juniper and ephedra on the north-facing slopes, but the hiking was warm in the sun. I was walking across the edge of Sinbad Country, and beyond the Knob from where I stood was the Head of Sinbad. I've never been able to find out who named these places. Though it sounds like something a knotty old uranium miner might have come up with. But the names seem to predate the discovery of pitchblende here. So it must have been someone else. Good names, though, that leap gracefully from my lips in this wonderland of stone buttresses and towers and canyons the color of flesh and blood and bone, the color of the sea at night, and the color of the down that limns the arms and legs of women dressed in billowing pants and pastel veils. Good names, well placed.

Lots of tracks marked the remaining snow, but all other animal life was elsewhere today. Differing water needs, as I said. Because some of my best friends are plants, or nearly so, I've no complaint about walking without animals, so long as the sun shines.

The doctors in their short white jackets told my mother not to take too much hope from the healing. True, it was likely now that I would survive. But it was likely, too, that the scar tissue that was forming across my wounds would cause me to be stooped, curl my gait, and make walking painful, if not impossible. They felt she should know that.

It might have been. Who knows? But about a week before I was burned my father had too much to drink, a lot too much, and spent the better part of our grocery allowance for that month and the next on a tricycle. A candy-apple red one with white streamers in the handle grips and chrome spokes in all the wheels, black-and-white rubber tires, and a bell that rang like the ice cream truck's. After the scalding I had to stretch some to reach the pedals, but with a little effort I could just make it.

I'm not much of a climber anymore, too much exposure, so to speak. But I managed to scramble a ways up the Knob. There is no more powerful force in all of nature than the force of water. It was water that chiseled this knob out of the Entrada bedrock, chipped it, bit by bit, from the bone of the earth, and left it here for me today. And it was water, expanding and contracting, flowing and standing, that powdered the first set of Rocky Mountains and left their remnants strewn all over Utah. Water, too, that carved out Devil's Canyon and the Chute of Muddy Creek, the Dirty Devil Gorge and Glen Canyon, Marble Canyon, the Grand Canyon. Water that laid the delta flat as the fist of God at the head of the Sea of Cortez. And just then, water—pushed through the collagenous pipes of my trunk and limbs, my neck and head, pushed a little faster after the climb up the side of the San Rafael Knob—mostly water that warmed me in the cold wind that raked those rocks today.

To the doctors, though, it appeared that the scar tissue was continuing to grow on its own, past its protective limits, continuing to grow progressively, like a tumor. They called them "keloid scars"—words that still make my stomach twist and my jaws tighten—and they recommended radiation to stop the excessive growth. I remember the cold, lead cup they laidover my penis and testicles, and the gray and green nozzle of the machine that belched out X-rays as easily as dragons belch fire. I remember it clearly.

The San Rafael Knob atop the San Rafael Swell

The view from the San Rafael Knob is . . . complex. There is great beauty, of course, but there is more. To the north are the four flat lanes of I–70 and the Eagle Canyon Bridge. There's a model of that bridge in the John Wesley Powell Museum in Green River. And just across the hall from the bridge there's a one-armed mannequin dressed to look like Powell himself as he sat in his wooden chair atop the dory *Emma Dean*. I've no idea whether it's a good likeness. But the display makes crystal clear the audacity of Powell's claims that even without the arm he lost at Shiloh he personally would map the canyons of the Green and Colorado Rivers. If you find yourself in Green River when the weather's too poor for hiking, take a look at that dory and stand for a moment next to that one-armed man.

Beyond the interstate is the fine firm fist of Chimney Rock and the elegant spires of Joe-and-His-Dog and Devil's Monument. And in the blue distance, the Book Cliffs and Cedar Mountain. To the east is Sinbad Country and a lacework of dirt roads rutted out from the precious little soil here by a near-continuous stream of off-road vehicles, a lacework somehow accepted, even encouraged at times, by the U.S. Department of the Interior and the Bureau of Land Management. Beyond that, the San Rafael Desert, the snowcapped purple peaks of the Manti La Sal, and the growing blister of Moab at their feet.

Toward the southeast, the tan finger of Turkey Tower and the mittened hand of Family Butte point to the miraculous blue sky. And then the iced pates of the Henry Mountains, and, in my mind, Bullfrog Basin and the watermarked walls of Glen Canyon.

It is a hard thing to see so much at once—so much beauty, so much desecration—and to try to hold it all in your eyes and your head, in your nose and on your tongue, hold it for later, when you know you will need it, all of it.

As I was finishing graduate school, as the last of the requirements for the Ph.D. in pathology were finding their way into my books and papers, the scar tissue finally broke down. The plastic surgeon said skin grafts would be necessary to replace the granuloma that had blossomed on my abdomen, just behind my belt buckle. The skin came from the fronts of my thighs. You can still see where. By the second week in the hospital bed the grafts were riddled with holes eaten there by staphylococci that refused to be concerned about penicillin. The surgeons just kept giving me more penicillin.

A friend, serving public health duty as an alternative to Vietnam, took the time the surgeons couldn't find for a routine culture-and-sensitivity test. The antibiotic in the IV was changed. The holes filled in, the seams closed up, and I published two papers on the structure of rat cell-surface proteins.

It was a short trip down the flank of the San Rafael Knob and back to the flesh-colored folds of Devil's Canyon. I walked down-canyon, thinking about the views across this wrecked section of central Utah, ruminating over the awful force of water. Gradually, I worked my way back down through the petrified layers of sediment. Jumping the short falls to the canyon floor below, watching the rim, thinking of 100 million years of occasional rain and ice and what it does to stone.

The final jump seemed longer than any of the short falls I remembered climbing on my way up-canyon. Quite a bit longer. But I have a tendency to do this to myself—to imagine, even when there is only one way I could have come, that nothing is familiar, that somehow I've gone wrong and that this is not the way I came in. I've come to expect this. So I walked on.

Abruptly, I was standing at the top of the seventy-five-foot, slightly overhung dry falls I'd seen from below. The walls of the canyon where I stood were steep and smooth as water-ice. No way out from here—I'd have to go back up-canyon. A rodent of memory began to gnaw at the back of my brain.

Twenty yards back up-canyon and I was at the last little falls I had jumped down. It looked less little from this side, a lot less little. The falls was nine or ten feet of underslung rock and vertical chimneyless stone shooting up to the rim of the canyon on either side. I stood on tiptoes and reached for the top of the vertical face. My fingers fell a good two or three feet short of the top. My stomach tightened, my mouth wadded up with cotton. I should never have been stupid enough to do this walking alone down an unknown canyon. I knew better. I've read those stories, I goddamn knew better! But I missed the stick I left in the sand to mark where I should've climbed out. I assumed I'd climbed this falls on my way up. So I jumped. But I sure as shit shouldn't have.

I took a couple of quick breaths and walked back to the big falls. I looked long and hard at the precipice itself and then the walls of the canyon. Nothing. Not a single rotten ledge even to work my way up to where I might have a chance of forcing some crevice to the rim. I walked back to the little falls. Panic snaked out from beneath some old rock at the base of my skull. Jesus, I wasn't even in the canyon I told my wife I would be in today! I jumped as high as I could, then jumped again. Nowhere close.

For a minute, maybe ten, I couldn't see anything except the lip of that little vertical strip of stone. Then, in a full panic, I tossed the ski pole I walk with to the top to free my hands, and I jumped two or three times, reaching for some handhold. Then, I though of how I could have used that ski pole, thought how stupid I was to have thrown it. Minutes more. My breathing quickened. Deep breaths, take deep breaths, reach, find a calm spot. Now, dammit!

For an instant my vision cleared and I looked again at what surrounded me. There were loose rocks everywhere on the canyon floor where I was standing—big ones, stackable ones.

A moment later I stood at the top of the little falls, gathered myself and my ski pole, and walked back to the stick I'd shoved in the sand to mark where I dropped back into the canyon. I scrambled up the steep scree to the rim. I sat then, to finish the process of reassembling myself. I took in and held a lot of air, I untied the knot between my shoulders, I thought about crying.

Three years ago an ugly blue-black spot appeared on my thigh near where the nozzle of the machine boiled off X-rays to slow what might have been progressively growing scar tissue. The doctor said it was a basal cell carcinoma, the one to have if you're going to have skin cancer. I'd have liked another option. He cut it out. A year later, he had to remove it again. Couple of stitches, no big deal. It hasn't come back since. But there's another place, not cancerous likely, a place where the scar tissue itself has thickened. At times it hurts some. But it'll wait, until after the New Year anyway.

Everyone else has left the bar. It's just me and Shirley now. She buys me another round.

"So why do you come here?" Shirley asks.

"To hike mostly, to be near the river, and to hike. So, why do you stay here?"

"I was born here," she says simply. "I don't know nothin' else. Besides, I like it, the emptiness, sunsets over the Swell. I like sloppin' beer and hustlin' burgers, and the boss is real good to me—time off when I need it, a free beer now and then—that sort of shit."

We share her cigarette for a bit.

"Only disadvantage," she says, "is that everybody round here knows exactly what everybody else is doing. Exactly!" She looks directly into my eyes.

"That'd be tough," I say.

"It ain't always so bad." She looks away, and stubs out her cigarette.

I thank her for the beers. She tosses back the last of hers.

"I gotta pee," she says, and heads for the back room.

I leave five bucks on the bar and walk out into a cold, fiercely moonlit night. Water creatures, all of us.

She Said Forests

"Can't you see?" She said, "Forests
aren't filled with trees, it's only the dead
pulled to their knees by nothing
more meaningful than a hint of rain."

"And bears," she said, "are just old leaves
held together with elk's hairs and
fish hooks. Stones
are goat noises that clatter

down walls of canyons
in Octobers and Decembers. And birds?
Only eyes
set in clumps of leaf-mold and duff

that we may see.
Only the water," she said,
"is real, and all, but only
all, that it appears to be."

His sickness was only part of something larger,
and his cure would be found only in something
great and inclusive of everything.
— Leslie Marmon Silko

In the Beginning

Every story must have a beginning. Whether or not a story has an ending is not so important. In fact, some of the best stories ever begun were completely spoiled by their endings. But a story must have a beginning. And though much has already happened, this is the beginning of this story.

La Poudre Pass, Colorado—the Beginning

It is morning, and I am standing on a knife edge between two of the greatest drainages in all of the world. Behind me, every drop of water that falls as rain, snow, or sleet is gathered by the Mississippi River and shipped to the Gulf of Mexico. Ahead of me, each teardrop, each hailstone, each spilt gin and tonic gathers eventually in the Colorado River and flows to the Sea of Cortez—a glittering topaz, inexplicably named for one of the top ten, maybe top five, assholes of all time—Hernán Cortés. It is no accident that I am here.

This spot, this spot where I am standing 11,000 feet above the Gulf of Mexico and the Sea of Cortez, is called La Poudre Pass. It was named by French fur trappers anxious to recall where they'd left their gunpowder, or so the story goes. La Poudre Pass splits the Continental Divide at a most unusual spot. East and west from where I'm standing the Divide heads due south before it resumes the generally southeast-to-northwest line that it follows in this part of Colorado. So here at La Poudre Pass as nowhere else, the Divide forms a horseshoe, a cup. That cup is the Kuwaneeche Valley, and its rim is the Never Summer Mountains. Inside this cup a river is born. There is a grand sense of the continent here.

This morning La Poudre Pass is filled up with fir trees and Steller's jays, and a little fog here and there left over from last night's rain. The air is a syrup of columbine and chickadee, and the sky burns with a blue-black light. It seems that if I had eyes only slightly stronger, from here I could see beyond the far blue scarps of the Gore Range clear to the great western ocean.

To my dismay, there is a road that runs across La Poudre Pass, and for nearly a mile beyond the pass the trail (once the La Poudre Pass Trail, now the Colorado River Trail) follows that road. And next to the road there is a ditch. In that ditch, west of the divide, there is water flowing slowly east. East!

I know from the maps that this canal is the Grand Ditch. And I know there is another like it carved into the mountain's muscle on the opposite rim of this cup. The road was built for the creation and the maintenance of the Grand Ditch. The fact that the ditch is not very grand and the fact that the road is now inside Rocky Mountain National Park make no difference. In the West, water is serious business. Together, the Grand Ditch and its sister, the Specimen Ditch, steal about 30,000 acre-feet of water per year from western Colorado, southern Utah, northern Arizona, southern Nevada, eastern California, and Sonoran Mexico. Quite a feat. But for the water-poor cities of Colorado's eastern plains, and for the politicians in those cities, the ditch is vital. So the river—even before it is born—is bled.

There is a stark mixture of the primitive and the practical here, a vivid picture of both the high-mountain riparian wilderness and the lengths to which humans must go to slake their thirst. There is a smell of the ocean here and a deep sense of the past. There is also a taste of the future, a taste of salt.

In spite of it all, or perhaps because of it all, it is a beautiful morning, and it is impossible for me to ignore the pull of the mountains, my proximity to the sun, the imminent birth of the river, the slow roll of the planet. And finally, to my delight, the trail leaves the road. Just before it does, there is a first breathtaking view of the Kuwaneeche Valley and the Never Summer Mountains. And in the center of this stone crib, wrapped in swaddling, is the infant river.

I have never been midwife to a river. And I'm not sure I will know what to do when the time comes. But there's no longer time to worry about that. Quickly the trail drops into the trees, and the river disappears from sight.

As the trail descends from the Divide there are a thousand little seeps, a hundred falls, a dozen twisting rills—pieces of river—working their way purposefully downhill. A Steller's jay shouts something obscene at me from the firs overhead. And though the sun acts like nothing is happening here, I know better.

Something miraculous is about to happen. Beneath my feet a river is gathering—right here, right now. Western America's greatest river, the river that has moved nearly half of Utah and Arizona into Mexico, the river that spins the turbines beneath Powell and Mead, the river that floats the Cigarette Boats on the Parker Strip, the river that flushes the 10 million toilets of Los Angeles, the river that changes the Imperial Valley from desert to Eden, the river that washes the last filth of western America into the Salton Sea is gathering between my toes. Never mind that it wasn't called Colorado north of Spanish Bottom in Utah until 1921, never mind those who would argue that any one of a thousand other streams might have been chosen Colorado. One look at the map, one look at where the contour lines dip their fingers into this cup, and you will see that this is the river's womb.

And today, no one but me seems to care. Fine. Actually, nearly perfect. It is probably best to be alone when making such a fool of oneself over such a simple blessing. I have never before had this sense of gravity, this picture of the tilt of the continent, this sense of what a purposeful being water is.

I discover that I have stopped walking, that I am just standing and muttering. Grateful that no one is watching, I force myself to once again begin putting one foot in front of the other, and I follow the trail down—down toward the sea.

Near the end of this first descent, the trail moves out onto the rim of the Little Yellowstone Canyon, the first major defile carved by the

Colorado River. The canyon has grown where the river has pared a limestone wedge from the spine of the Divide itself. And the gorge bears clear testament to the power of this barely born river, and the canyon speaks in the scrawl of its limestone runes of what is to come. Already the river is reshaping the continent.

Beyond the Little Yellowstone, the trail and the river spill out into the Kawuneeche Valley. Though surely carved by glaciers, the valley is indelibly marked by the river—the deepening cleft, the old meanders, the lush meadow. Here, the river has grown to eight, in some places twelve, feet in width. It is lined by low-growing willows and sedges. The ramp of the Never Summer Mountains rises to 12,000-foot summits on the west, and the massif of Specimen Mountain pushes up the sky to the east. A place as magical as the words that name it.

The trail leaves the river and moves off into the trees on the eastern side of the valley. For a while, I follow the trail.

Water is a remarkable thing. Every drop of it was formed in the forge of a dying star, a supernova. And the water carving its way through this valley is older than any living thing on this planet. Water has been protoplasm, it has been rain. It has been clouds, it has been comets. It has been urine, semen, sap, seawater, swamp water, sweat, tears, nectar, blood, pus, wine, sun dogs, moon rings, rainbows, ice floes, ice plant, icicles, and eyes. It has been human, animal, vegetable, and mineral. It has been a thought, a kiss, a word spoken into the frost of a winter night. It has been the tongue that spoke that word. It has been the ear the word fell upon. It has been the frost. It has been the winter night.

The path that brought this water here is unimaginable. The force that gathers this water and makes it into a river is beyond human thought. Not only, to paraphrase J. B. S. Haldane, more complex than we imagine, more complex than we *can* imagine. And that ol' river he don't say nothin'. He jes keep rollin' along. This water has made me a child again.

I leave the trail and the trees and walk to the center of the valley. I wish to be near the river. This morning I came here to escape. I'm not sure from what—something, though, that has been eating on me for some time. It was a good choice, coming to this spot. For a while, whatever drove me here leaves me, and I sit fashioning stick boats from fir and spruce twigs.

There was a city here once, right here where I'm sitting. Lulu City, a community of nearly two hundred men and women and children, chickens, dogs, mules, houses, hotels, and brothels. One hundred and twelve years ago there was a community of silver miners here. Nothing is here now, nothing to mark this space touched by the dreams of so many people. Nothing, except the river patiently scraping at the rocks, pulling up the soil—just as it was 112 years ago, just as it will be when I die, when my children die, when their children die.

Once I've assembled a respectable armada, I begin setting my ships to sail on the nascent Colorado. Overhead, clouds form and dissipate in the deep blue of the 12,000-foot sky. Form again, dissipate again—over and over, like tasseled anemones. My flotilla launched, I watch as the wooden boats rise with the small waves and crash into the infant troughs. I watch until each boat has passed from sight. And then I am jealous of whatever beetles might have bored their way aboard, any flies that booked passage, of the mold growing inside the wood.

From where I sit, the Colorado reaches nearly 1,500 miles to (in wet years anyway) the Gulf of California. The river passes through canyons that stretch the imagination of even the most imaginative among us. This river has been fought over by seven states and two nations, and not one of the great cities of the American West could survive without it. The river once carried great tidal bores into Arizona and paddle-wheelers as far north as Parker. All of that is gone, of course. But today I have what remains to myself. I envy the small craft I have loosed. I would like to see all that they will see of this river.

For a while I rest beside the young water. I dream of horses and shorebirds, dream of these creatures in no way that seems useful while I'm awake.

When I rise, I wish to do something for the river, something to show my gratitude for this day and these images. But what do I have that a river would want? And then as I think for a last time of the boats I have sent to the sea, I know. I move to the river and stand in the thin flow near the shore. I unzip my shorts and begin to pee into the Colorado River. As I am standing, thus exposed, a miracle happens. From the Sea of Cortez, no farther, from the Pacific Ocean into Mexico, Arizona, California, into Nevada, into Utah, into Colorado, through the Black Canyon, the Grand Canyon, Glen Canyon, through Cataract and Westwater Canyons, through the Gore Canyon, through the Kawuneeche Valley, up the glittering shaft of my urine, into my penis, into my ureters, my kidneys, my blood, my mind, the river comes. And for an instant there is only a single thing, a solitary soul stretching from Pacific to neuron, one thought.

For this instant, I am afraid. I am afraid of the picture in my mind, I am afraid of this powerful river, I am afraid that I will not be able to get my penis back into my pants and that I will have to drag the great length of this penis behind me for the remainder of my days. But as the flow of my urine diminishes, the vision fades, my penis returns to normal size, and I am able to stuff it back into my hiking shorts. I am stunned.

I stand with the Colorado running over the toes of my boots. Thunderclouds are gathering overhead. I should be heading back. But for a moment I can't. I can't do anything but think of the river. I cannot shake the sense that I have left a door ajar, a door that will never close.

I begin moving my feet, I begin thinking of home. The climb to the Divide is filled with the scents of lupine and pine, the songs (if you want to call them that) of jays and nutcrackers, a faint taste of cinnamon. As I walk, it begins to rain. And I realize I am speaking to myself. Some of the words are the river's words, some of the words are God's words, some of the words might be either.

The Snag at Potato Bottom

The great rocks
stand for all things.
For the marks of their sediments

and the motions
of the sacred seas
are in all things,

things animate
and things sessile, things
sentient and things dead.

And at the center,
buried in the fist of the rock,
is the river,

the river that is all things,
for all things blossom from water
all things seek to hold water—

carrying its waves
and its precious silt
as though it still mattered.

And it was the wolf
of time
that carved these rocks,

that muddied this water,
and then, in his shame,
buried them

within each of us,
beneath the fragrance of the sage
and the fierce red coil of the wind.

Maybe no one can distinguish which voice
Is god's voice sounding in the summer dusk
Because he calls with the same rising frequency,
The same rasp and rattling rustle the cicadas use
As they cling to the high leaves in the glowing
Dust of the oaks.
—Patiann Rogers

Desert Sonata

Michael is sorting stones. Sometimes he prefers them that way—sorted, that is. I'm sitting in a low-slung chair humming a Merle Haggard tune I can't remember the name of, and somewhere a raven is playing tenor sax. Bull Canyon, a fine fissure in the Colorado Plateau, gathers in blood reds and chocolate browns just beyond my boot tips and rolls off toward Westwater Canyon and the Colorado River.

Michael and I have come to wash the city off our skins, to breathe air tainted only by the songs of birds and the light of the sun, and we

The author's brother, Michael, and his favorite form of transportation, near Kanab, Utah

 have come to walk in the desert until we have lost our love of things. But that can wait.

Dinner is done. Too early to think about dessert, so I hum, the raven plays a riff, and Michael watches it all with his careful blue eyes. So slowly you might imagine time really doesn't matter, the afternoon gathers up the blue robes of evening and walks off to the west.

Michael brings wood then, and builds a small fire among the stones he's been rearranging. The flame he kindles flickers in his eyes, dances among the wrinkles in his face, and skitters out into the deepening dusk. My brother has grown nearly old—his blond hair mostly gray, the crow's-feet deeper, only the blue in his eyes fierce as ever—grown nearly old while I was looking elsewhere. What bothers me most about this observation is that it leads to one inescapable conclusion—I have grown nearly old as well. We've too few of these trips left to us.

The air fills with a piñon-smoke marmalade and shadows like eels swim up from the canyon. Michael and I, then the piñons and the junipers, draw close to my brother's small flame. A poor-will calls out, once, twice, . . . once more. Michael stretches his thin frame before the fire, pops a vertebra or two, and then settles with me among the shadows for the serious work of fire watching.

For a while we talk. Michael tells me stories of trips downriver, downcanyon, stories of Skull Rapid, scorpions, and quarts of tequila. I talk of hiking and climbing in Canada and fishing with my son in Mexico. But the talk adds up to little, and the evening seems too fine for idle words. So as the edge of the galaxy spirals out across the sky, we fall into a final silence. And as though this were some sort of signal, the wind begins to blow softly and the trees step fully forward, the piñons, the junipers, and begin whispering stories of their own.

And they are still whispering among themselves when our fire, like a broken ruby, collapses into itself and the Northern Cross falls from the sky. There is a sound a fire makes then—when it is finished for the night—a final ashen sigh. Much like the one you know your skin and bones will make when you give up your last breath. I manage little sleep amid all the whispering, but Michael, having heard many of these stories before, sleeps well.

Next morning I wake around seven, but Michael is already about. Coffee is boiling on the stove and bacon and eggs and buttered bagels fill a plate near my nose. God, I love being waited on. I eat in my underwear while the sky collects itself and the sage and the juniper paint the morning air.

Morning toilet, a pair of old shorts, and Michael and I head off for that walk we promised ourselves. We are near the easternmost edge of the Colorado Plateau. The canyons here head mostly westward through some of the first sediments flensed from the buckled earth by the Colorado. It is an old place, where many things have happened—some seen by humans, some not.

We follow a game trail out onto a thin scarp, then drop quickly through some rotten sandstone to a trackless waste of piñon and juniper, blossoming prickly pear cactus, and bunchgrass. The canyon drops once more, as canyons will, through a second layer of sediment to another plateau, flatter and broader than the first. It is already warm, and the air has begun to wrinkle above the stone shelves.

Lots of sheep sign here, desert bighorn—a mammal I've never seen outside of books. Michael says he's seen sheep once before, but he can't recall where (a ranch in New Mexico I surmise, but keep it to myself). Down a water-polished, stone-bellied wash we work our way toward the river. Walking steadily and quietly, we imagine a cold river-water bath at the end of all this.

And then, abruptly, the canyon we're following rims out at the edge of a much deeper and narrower canyon. Three hundred dead-vertical feet of Wingate sandstone, then a ledge of Chinle, and somewhere beyond that, beyond sight, another canyon, then another, and then, perhaps, the river. The black-and-orange Wingate formation tends to fracture in great columnar chunks, leaving vast dihedrals and lots of warm air. We work our way along the rim in both directions, hopeful—nothing but perpendicular rock, black with varnish and empty of promise. The end. No river today.

We fuss and grumble for a while, but our frustrations melt quickly in the warm desert air. We spend the rest of the morning aiding boulders on their long journeys to the sea, herding sage and Spanish bayonet along the stone-bottomed arroyos, chasing lizards, humming to ourselves. There's no accounting for it.

We walk back for lunch—more bagels, some cheese, a little summer sausage, then a Snickers, for me. Michael brought apples and bananas, thin crackers, and Gatorade. This lack of fat in his diet will get him in the end, I'm confident it will. Already he's five years older and much grayer than I am. I take a moment to remind him. He thanks me.

Over ninety degrees now, good hiking weather. We begin again as cumulus blimps blossom overhead. This time we choose the higher course, the narrow walkway across the Entrada rimrock. This route goes, not to the river of course, but out across scarps of ancient rock pitted by wind and ice. Here and there a little soil has collected in the stone cups of the Entrada formation, and much of it is colonized by cryptobiota, soil like blackened oatmeal, thick with microscopic creatures slowly changing stone into trees—protist wizardry. Lots of lizards—finger-length whiptails, chunky collared lizards, and others I can't name. Stoic remnants of the Cretaceous, still pissed, near as I can tell from their scowls, that they missed their shot at birddom.

Piñon and scrub jays, common titmice, move bulletlike from tree to tree, the jays screaming at us and at one another. White-throated swifts and tree swallows sweep insects from the air, spiraling out over the depths, and raggedy-assed ravens shout at things only they see. Desert sonata.

The last of the game trails peters out at a little neck of stone sticking halfway into the void above the canyon and ends in a broad pulpit of buff-colored Entrada sandstone. From there a paper-thin arch, no more than three feet wide, leaps a gaping pothole thirty or forty feet across. Nothing but wind below for half a football field or more.

Michael is staring at the arch. It's maybe three feet through at its thickest. But I imagine I know what he's thinking, and hope I'm wrong. Grand Mesa, snowcapped, rises stoically to the east, and a watermelon-green slice of the Colorado, Moab-bound, spirals past in the distance. I sit, pull a bottle from my pack, and take deep drafts of the warm water. Michael walks to the far end of the arch. He looks for a long time, then comes and sits beside me. Though nothing's been said, my stomach tightens with adrenaline. We sit, not speaking, both of us acting as though nothing is happening.

Then Michael lays his straw cowboy hat beside me and walks to the near end of the sandstone span. He hesitates a moment where the crumbling end of the arch meets the main column of rock, then he simply walks out onto the arch, hunkered a little as though this could help break his fall when the rock collapses beneath his weight. Three, four steps, then he straightens and the wind gusts against him, pulls him a little toward the near side. I wish I could turn away. He staggers once, then regains his balance. Another step, surer, another, then he straightens fully, and his strides lengthen. Looking at nothing but the stone, he places one foot in front of the other and moves quickly across. Then abruptly he's back beside me, grinning, breathing in gulps of hot air. My palms are oily wet.

"Jesus!" he says in a harsh whisper. "The whole time, the only thing I could think of was, I hope I have the presence of mind to just turn and shrug if the whole fucking thing lets go. Problem was, it was too damned narrow, you could see the drop on both sides of your feet all the time! That was something, really goddamn something! A walk in the goddamn sky." Then he lapses again into deep breaths and tries to wet his lips.

There is a thing among men. I've no idea where it came from, something primordial, maybe limbic. So now I'm eyeing the arch, measuring its width and the long airy run of the delicate span. My stomach tightens one more notch, my mouth fills up with ashes. I drink from my water bottle and wrestle with myself.

Nope. Not today.

A long time ago, on a relatively simple climb—a climb I'd led before—I panicked on lead. Panicked so thoroughly I couldn't move—up or down. I simply hung there, drawing great ragged lungfuls of air and pumping bolts of adrenaline. The people below couldn't help me without considerable risk to themselves. So for maybe five minutes (seemed like forty-five), we all just clung to the rock wondering what the hell I was going to do. Finally I slowed my breathing, took tenuous hold of myself, and slammed every piece of protection I had into the rock, every nut, every stopper, for an anchor. Then I rapelled off the face of that cliff, leaving behind about forty-five dollars worth of hexcentrics, carabiners, and nylon webbing.

Once you've done something like that you're never quite the same. Once you know you're capable of that level of insanity, of that depth of despair, of placing those with you in that sort of danger, the John Wayne bullshit just doesn't work anymore.

I take a last glance at the sandstone arch, as beautiful and delicate as a woman's wrist, and I gather my stuff and start back. Michael doesn't say a word, doesn't care I'm sure, what I chose to do or why. But I'm pissed, pissed that once again I can't leave this bit of black baggage in some godforsaken bus stop somewhere.

As we walk, I rag on myself for missed opportunities, draw long pulls of air, fresh and dry as flint, and hiss at myself beneath my breath. After a half mile or so we stop to rest in the shade of a large juniper. Overhead, clouds form and vanish, fall into shapes of griffins and gargoyles, dragons and dinosaurs. A canyon wren spools its rich song into the warm air.

Michael pulls out a map, and with compass and altimeter he pinpoints our whereabouts among the brown whorls of the geodetic survey. I trace circles in the billion-year-old dirt at my feet, ignoring everything.

The problem is that sometimes that same panic shows up in other places. It has dragged me out of my tent at night and off a bus or two in Mexico. And just the possibility of panic out in the middle of nowhere has made me back out of two trips with Michael. The level of commitment was just too great, and I regret almost daily not having made those trips. How much longer will we have? Were those our last chances for treks into the Grand Canyon or across the great mesa of the White Rim?

But when it comes for me, there is no reasoning with it. And it's hard to accept that a part of me is totally beyond my control, and always will be—small, I imagine, the color of cave worms. But mine, and me, no matter what I'd like to think.

I heave one more rock into the canyon and then shoulder my pack. It is hot enough for the walk back now. We choose again the path nearest the edge, the one with the view. I ignore it. Michael and I talk about my panic sometimes, but it's hard to explain in a way that's touchable by someone who's never felt its scaled feet hooked in his own scalp. In spite of that, Michael tries to understand, and even when he doesn't understand, he accepts it. It pisses him off when it ruins our plans, but he accepts it.

Occasionally we stop while Michael helps a stone on its way down, otherwise we move steadily back toward camp. Halfway back, Michael works down into a broken waterway that looks to me to lead to nothing but air. But it ends, instead, in a stone grotto, shaded since morning. Deep cool. Ravens and vultures float on bubbles of hot air rising from the canyon. This day I had imagined a careless walk to the river. The river imagined otherwise. I work to accept that. Nearly chilled in spite of the heat, we move back into the fist of the sun and finish our walk.

In camp that evening we've drawn our chairs to the rim. Pulling hard on cold beers, we watch the light paint the rock walls while we wait for evening to roll across the rim. I'm still grumbling inside when suddenly six desert bighorn emerge from the brush and cross the rim below where we're sitting. Six white-assed, black-eyed bighorn sheep—still shedding remnants of their ratty winter coats—no more than fifty yards from where we sit. Clattering across the rimrock, each stops briefly to consider us. Convinced, and rightly so, that two such aged and out-of-shape humans pose no threat, the rams and ewes clamber slowly off downslope, seeking on their needle-sure feet the deep shade and water-rich potholes of the lower canyon.

For a long moment we stare at the spot where the sheep materialized from the piñon-shaded stone. Michael turns and smiles, sheepishly I think. I return the gesture, best I can. Then, in place of conversation, we each pull our chairs a little farther out onto the rim. I open another beer, and Michael returns to sorting stones. I sip at the beer, and the orchestra within—the one that's been playing one Russian composer after another all afternoon—slows, then stops. The sky slips on a blue-black negligee, and the stars come out to see what I will do with all the things this world has given me.

The Grand Valley from the mesas near Bull Canyon

Near Gypsum Gap

From where I stood, when the light fell
at last among the furrows of that distant
headland, I thought: You and I

could go there and live
beneath the alligator junipers, the fluted
sky, and the high, hard shoulder

of the land. And there,
there would only be the promise.
Rain and wind and ruined rocks,

sunsets, starlight, the dry chill
of early mornings. We could live
there, in the shade of those trees for

a very long time, years
perhaps. Until one night, satisfied
with it all, we would purposely

peel our fingers loose
from the grasses and the hollow
stones and fall into the stars.

"How far could you swim, Daddy
in such a storm?"
"As far as was needed," I said,
and as I talked, I swam.
—William Stafford

Grand View Point

There is no sound save one—the winter rales of the Colorado River, ice against ice. Frozen floes grinding slowly south past iced mud and gravel. The rest is snow and fog and a blunt, wet silence. Even the birds are elsewhere today.

I've been squatting at the river's edge for twenty, maybe thirty minutes. My legs are stiff, my hands and feet are numb, and about all of this the river doesn't give one shit. For a few minutes longer I watch, through the cloud of my breath, the Colorado shipping ice from the snowfields of the Never Summer and Gore Mountains to the warm thin streams of the desert southwest and the improbable green of the Imperial and Coachella Valleys. Ice, at last, for margaritas and gin fizzes sipped by sunburned people beneath striped canvases while the sea curls over upon itself over and over again.

Patrick is smarter. He waits in the warmth of the motel room in Moab, watching morning television until I've had my fill of the river. Though twenty-three of my chromosomes are ratcheting out protein inside of Patrick, his water need differs considerably from mine. But that may change. I have.

Patrick and I have been here before—twice, in fact. The first time with Nancy, his mother and, back then, my wife. But I doubt he remembers. Patrick was three years old then, his mother young and dark, beautiful as obsidian, but flawed. The second time was a year ago last October—not quite twelve months after Nancy put a plastic hose on the tailpipe of her Subaru and filled her mind with cloud pictures. Not quite twelve months after Patrick and his sister had to choose to unplug the machine that forced their mother to breathe after she no longer cared to. It's an odd damned world.

That October, when we were last here, summer clung to the year more tenaciously than usual. The leaves were still mostly green, the river deep brown, and the afternoons were cream over warm cobbler. But we had only two days before he would leave me again.

Patrick was on his way to California. I'd driven partway with him to find what was left after his mother had taken what she could of him. We arrived in separate cars. At the motel, I told the beehived woman, the woman wearing the emblem of Utah on her head, that I'd take the room for two nights, hoping Patrick would stay the extra day. She passed me the forms, I described someone's car and gave her someone's license number. I don't do this to deceive; accuracy at times such as these just doesn't seem worth the effort.

"Have a nice day."

"You too," I said, with the same sincerity she'd mustered—that is, if one of us were to be eaten by rabid javelinas later in the day, it wouldn't make much of a difference—to either of us.

Patrick and I carried our things into the room. My clothes were wadded into an old blue nylon bag marked by days in the back of my pickup. Patrick carried an olive drab duffel that had been his mother's while she served with a surgical unit of the army reserves. Her name still stood in black block letters across the side. And Patrick still drove the car she'd driven, the car she was in when she'd run out of reasons.

Patrick dumped his bag at the foot of the bed nearest the window and opened a beer. I followed his lead. For a bit we sat drinking the cold Millers, trying to get the wheels out from under our feet. I turned on the TV like I always do in motel rooms. We stared at it awhile, not watching. Instead, I watched Patrick when I could, hoping his nineteen-year-old body and deep brown eyes would give me some sign of what remained, some hint about what I should do. He had her hands, long slim fingers, and her neck, but the deep brown eyes and the broad shoulders were mine, so was the firm set of his chin beneath its dark stubble. But that was all, just bits of the past. Nothing more, nothing I could use.

After a couple of beers I got interested in being in the car again. I suggested we take a drive out past the potash plant and see some of the White Rim. Patrick agreed without enthusiasm, pulled on his tennis shoes, and headed for the car.

Beyond the potash plant and the end of Utah 279 the road surface changes character abruptly. Even the river abandons the road here and heads south. I'd say the road turns to dirt, but there is precious little dirt here. We jounced and jerked our way across bedrock until we encountered our first major hill. I'd been over this hill before and since—it's really not much compared to legitimate hills like the Murphy Hogback or the climb from Potato Bottom, but neither is it trivial. Patrick balked.

"You can go on if you want. I'll wait here Ger." He's never called me by anything other than my first name.

"Pat, it's okay. I've climbed this hill before, and the truck can handle it easily."

"I'd just as soon go back."

". . . Okay."

I turned the truck around, careful of the deep sand beyond the edges of the road. The glowering red and brown walls marked our progress but showed no sympathy.

Patrick had forgotten that he too had come this way—long before, down from the island above on the Schaffer Trail, a road much more exposed, much harder on sphincters than this one. He'd come down that road with me, his sister, and his mother, come down it more than once. Today, we turned around.

We crossed back over the rusted and broken spine of the land, back toward Moab. The red wounds of old sluice channels lattice the earth here—shallow ditches carved to carry potash sludge to the evaporation ponds below. And then the garish insult of the plastic-lined ponds themselves, the hoses, the cables, the lavatory-green potash plant—finally the wall as we crossed back onto the pavement, the wall between Patrick and me, solid as ever. The evening turned a muddy brown and the sun crept off toward California. We drove the last dark miles in silence.

But the shattered afternoon had given one gift. Patrick had agreed to stay the extra night. So tomorrow we could waltz once more with what hung between us. This pleased me, and as always the river, the rocks, warmed me.

Dry cheeseburgers and rancid late-night TV gave way to the fine rime of a fall morning in eastern Utah. I suggested another drive. Patrick agreed even more hesitantly than before, and we drove off toward Arches National Park. Where we crossed, the Colorado was busy with mud and ducks and making those special curled waves it likes so much, and the water either didn't care or, more likely, was too preoccupied to notice me and my silent cry for help. Patrick, for no apparent reason, seemed less self-contained this morning, more animated. We talked as we drove, but spoke of nothing.

Once in the park, I pushed him from the car for photographs. First, at the prow of the Three Gossips and later beneath the precarious bulk

Dead Horse Point and the potash evaporation ponds, Dead Horse Point State Park, Utah

of Balanced Rock. Later, I found I was too far from Patrick when I took these pictures, and the prints are dominated by the red-and-black storm of the rocks. Patrick looks small and fragile beneath them and is completely unrecognizable except to me.

I suggested, with little hope, that we take a shot at the trail to Delicate Arch. Patrick agreed. I held my surprise just below the level I thought he might detect and drove on. Patrick likes hiking about as much as I like Republicans and committee meetings. At the turnoff to Salt Wash we left the paved road, and to my delight there were no other cars in the parking lot at the trailhead.

The trail to Delicate Arch jumps Salt Wash via a suspension bridge fine as a maiden's calf and quickly twists off into the slickrock. If you're down, or you don't know why it is that people like to hike or camp in the desert, if your gal has left you or you've just fallen in love, if you're depressed or manic, I recommend the walk to Delicate Arch, especially in the fall. The trail works its way over great humpbacks of fossilized sand, white and brown like human flesh, red like tomatoes, brown like worn wood. There are glimpses of dark stone fins to the north and dry washes and old canyons that speak of the ferocity of the infrequent rain and constant wind. Narrow channels curling off into the petrified marrow of the earth, rippled and wind-pocked stone walkways underfoot, then a brisk climb. And finally, out of nothingness, Delicate Arch. Its heart cut clean away by the scythe of the wind, and where its heart once was there is now the white pate and deep green scowl of the far-off Manti La Sal—Tukuhnikivats, Horse Mountain, Mann's Peak, et al.

We walked, and we emptied some. For the first time in a while the cloud between us thinned and we spoke easily. Not of great things, but easily, of baseball and football, Patrick's plans in California and Nevada, the day, the day before. I struggled to maintain the delicate contact, keep the small artery open, and avoid words that would clog it. The walk, Patrick, the slickrock, and the razor-sharp fall air were briefly wondrous.

But, back at the car the weight of the day and things yet unresolved fell upon us again, and we drove off with a familiar silence sitting between us.

Patrick kept his mother's death around his neck like a stone—old lava with veins of obsidian, bearded with moss. And we couldn't talk about it, even indirectly. But stone or no, Patrick hadn't stopped caring for his mother. Not even during our divorce and the messy custody suit that later brought him to live with me—caring, in the awkward way that a boy cares for a woman he must love but should, at times, hate. After her death, because there was no one else alive he could trust with his anger, he brought it to me. Like all of us, what he really hates is the way things are, the way things that look so goddamned perfect so quickly and so often turn to shit. But Patrick imagines I'm at the controls, and he hates that. But I'm also the only thing he has left to love—"It's a hard life, it's a hard life, it's a very hard life. . . " And you know what pisses me off the most? There's no reason it has to be like that, it just fucking is, that's all—it just fucking is.

We stopped for a late breakfast. In those days Moab wasn't known for its cuisine. My toast and coffee were cold, and my eggs looked and tasted like broiled mucilage. Patrick was reading the sports section of the *Denver Post*. I was desperate, I knew my time was running out. So, because I could think of nothing else, I suggested we drive to Grand View Point. And because Patrick couldn't think of an excuse good enough, quickly enough, he agreed to go.

About the time we crossed The Neck, about where you see that the only way out is the way you came in, about where you see that way is narrow and rotten, I remembered when I was here last—here with Patrick, his sister Jennifer, and their mother. It was the last of several trips we made to Canyonlands. No paved roads then, no fences, just two people and their children without a clue as to what lay beyond next week or the week after that, without a reason to believe it mattered.

I told Patrick. He'd forgotten, but seemed curious. I told him how his mother and I watched as a friend's dog nearly jumped the stone wall at Dead Horse Point, unaware of the thousand or so feet of down beyond that wall. I showed him where we all had camped, the four of us in our green microbus, our only company the red dirt and pale green sage. Told him how I slept across the door of the old Volkswagen so that neither he nor his sister could slip out at night and wander off toward the emptiness that lay in all directions save one. I told him of the campfires and stars, the hikes across the desert floor with him in the pack and his sister on my shoulders. He listened. I talked. The sky turned a deeper blue.

Where the slender finger of Grand View Point meets the empty sky of Utah, where the Island in the Sky ends, we parked and walked the gravel path to the overlook. Because it was October, we were alone. The White Rim stretched like snow beneath the awful dihedrals of the shattered Wingate sandstone. Bits of the river canyon, the Needles, the Six-Shooter Peaks, and, in my imagination, pieces of the Abajos, the Sonoran Desert, the Gulf of California, and my only son were disappearing into the blue soup of evening.

We stood as darkness gathered. I thought of the days here with Patrick and his mother. I wanted to say something, but everything that wrestled for space in my mouth seemed stupid or careless. My thoughts finally settled with the ravens and vultures in the thermals. But I knew that Patrick would leave in the morning, likely leave before there was another opportunity. So I finally said the first thing that fell out of my head.

"I imagine the spirit of your mother in places like this, Pat. I imagine it joyful." Patrick said nothing. He stood and stared at the void as the wind worked his fine brown hair and a scrub jay somewhere back in the junipers screamed at something irrelevant, or not. I stared at the

Grand View Point, near Moab, Utah

scars in the old rocks below and wished I'd thought of something else to say, something more, something better. The darkness thickened.

The next morning we shared a newspaper and breakfast, talked a little about the football scores, then Patrick got ready to leave. We walked to his car. I insisted on holding him for a moment. He handled it, stiffly. The breadth of his shoulders, the roughness of his beard frightened me, reminded me of holding my father. The day brightened as though none of this mattered.

"I enjoyed this time Patrick."

"So did I."

"I'd like to do it again sometime."

"I would too, . . . I know you're trying, Ger. I appreciate it, honestly."

"I love you."

"You too."

He left. I watched until his car—his mother's car—was out of sight. Then I got into my truck and drove. When finally I took the time to look around, I found myself back out toward the potash plant again, not yet ready for the ride home.

I watched the river on my left, the rock wall to my right. I passed two climbers, one clinging to fragile flakes of sandstone fifty feet above the pavement, the other in an armchair belay, watching completely. I stopped at a pullout I'd noticed on the drive with Patrick. There was a small canyon, its mouth clogged with undergrowth, leading off into the red rock. It seemed a good place to be for a while. I changed into my boots, shouldered my pack and water bottle, and walked up-canyon. The first hundred feet or so were thick with tamarisk and willow. It didn't appear to offer much and kept threatening to end soon, but it was all I had just now.

The canyon cut deeply into the Entrada sandstone and wandered much farther than I had imagined or was willing to go that day. I will find its end another time. This day, a mile or two in, I climbed to a spot above a dry falls that ended in a deep green pool of old rainwater, water that tasted of rock and sand and age, cold this October morning. I sat above the pond in a field of blackened stones, all that was left of some other rock layer that once stood above where I now sat. I lifted a palm-sized one of these, and I held it between both hands. The cold of the stone seeped through my fingers and palms into my arms.

I thought of Patrick's mother and the black gift she'd left him. I thought of how that gift felt to each of us who had loved her, how it had stretched the web that held us. I thought of the shitty job I'd done parenting. I wallowed in it all for a while. Then the coldness of the stone in my palms reached to my chest. My mind emptied. I looked at nothing. Thought deeply about nothing.

A year and a half later, the dry breath of the iced river holds me for a moment longer, then the flatulence of a tractor-trailer rig coming down from Crescent Junction drives me from the water. Patrick, now twenty-one years old, is waiting at the same motel. But today we are to drive to the Grand Canyon on the third day of a twenty-one-day trip to Cabo San Lucas, the length of the Colorado River. Driving back to the motel in my parka and gloves, it is hard to imagine the warmth of Mexico, hard to shake the depression of that October almost two years ago.

I gather up my son at the motel, and we head south toward Blanding. The day warms as we drive, and along the way Patrick and I take the turnoff to the Needles Overlook. At the rim we take the requisite photos of each of us standing beyond the signs warning us not to stand beyond the signs, and I look long across the White Rim to Grand View Point. It looks very different to me from this side.

That night, at Bright Angel Lodge, Patrick and I share a drink on a cold wooden bench 7,000 feet above the Sea of Cortez, 5,000 feet above the Colorado River, without a net. We talk freely, not about his mother, not yet, but we talk freely. Perhaps there is time: he is twenty-one, I am forty-five. Perhaps. But just now the iced vodka is good, my son is next to me, and the most god-awful canyon of them all is just beyond my toes.

The author's son, Patrick Callahan, walking on thin ice near Moab, Utah

The Way Things Are

"Are these stories very old?" he asked.
"Very
old." I replied. "Why
do you ask?"
"The stories speak of magical things: frogs
become princes, yarn spun to gold, evil spells, and the words
of wolves and foxes, ravens and dogs."
"They do indeed."
"Where has it gone?" he asked.
"Where has what gone?" I asked.
"The magic that these tales speak of."
"They are only
tales." I said. " Tales created
by people who knew little
about the ways of this world." And,
outside, snow began falling in great crystalline flakes.
"Then why this magic, these animals, these
golden rings?"
"Ignorance." I replied.
He began again to read, slowly
lifting words from page to mouth. And
outside the snow began forming little pockets
at the corners of the windows.
I lied to him,
Of course I did.
He is too young to know,
too young to hold,
such a simple truth.

I am circling around god, around the ancient tower,
and I have been circling for a thousand years
and I still do not know if I am a falcon or a storm
or a great song.
— Rainer Maria Rilke

Almost Like Dancing

I only do it with birds. I'm not sure why. Maybe it's that birds are the most nearly perfect things I know. Or maybe it's because birds are things I'm not—graceful, self-reliant, certain. It makes no difference. And I'd like to tell you I'm proud of what I do. I'm not. I just can't help myself.

I can't, anymore, leave birds—especially birds of prey—crumpled like Coors cans in the middle of a greasy highway.

So this morning, on my way toward Moab, at about seventy-five miles per hour, an owl—a big owl—face down in the far left lane, snares me. "Leave it lie," someone inside of me says. "It's the freeway for God's sake, and who knows when you'll get a chance to turn around again. Besides, it's dead!" But someone else inside has already moved the truck into the left lane, turned on the flashers, and started looking for a place to cross over. I find the next crossover only a couple of miles back.

And then I'm out of the truck, standing on the gritty shoulder of westbound I–70, a breeze full of river water searching in the bunch-grass. It's a great horned owl, the biggest one I've seen. And because even its eyes haven't yet been worked over by the magpies and ravens, it hasn't been dead for long. I nudge him with the toe of my right boot, not caring to go finger-first after an owl only partially dead. He doesn't move. I turn him over onto his back. Only once before have I had the opportunity to look at a big great-horned up close, and that one was already full of maggots. This one is like new. No blood, no broken feathers, no obvious injury at all. Good as the day God made him—except, of course, he's dead.

Dead or alive, though, this is a remarkable machine. One polished over millions of years by the steady hand of evolution, honed, near perfectly, for the sole purpose of killing. And by the size of this guy, he must have been pretty damned good at it. Toes tipped with hooks, big as my fingers—shining, this morning, like fire-polished onyx. A wrinkled palm, almost human—leathery, punctate, skin the color of hunger and death. And square in the center of his serious face a jet black beak, big as half my fist—a goblet for skin and scale, for flesh and bone. Covered over with feathers—like a blizzard of chocolate and sienna, cinnamon and teak, ivory and sea salt, moonlight and sunrise—no two alike. I spread one stiff wing to its full length—nearly a yard.

I imagine the owl in flight—slow hushed arcs beneath the darkness, eyes full of torchlight. I imagine the terror of shadowed field mice and cottontails, collared lizards, red racers, freshly born coyotes. I imagine a mouthful of warm blood and the sweet pop of gristle.

It's a good morning to still be breathing. I lift the bird into my arms and climb the near embankment. Above the roadway the land opens out to a warm sea of sage and pygmy juniper. Magpies are reading their morning matins, and in the distance the dark canyon of the Colorado splits the low ridge of the horizon. I walk for thirty yards or so, toward a low, thick-limbed juniper tree. The bird is heavy and warm in my arms. The other owl—the one full of maggots—weighed nothing, felt like an armload of winter evening. This bird has the heft of summer.

I lay the owl beneath the tree and prop its head so it faces southeast—where the sun will rise tomorrow. And then, I pull three of his primary flight feathers—long as my forearm—from the end of his left wing and slip them inside my shirt. It is essential. I try to think of something to say, a prayer maybe, something to say out loud. Nothing comes. I wish the owl good hunting and Godspeed.

I walk back to my car. The sun's breath is warm and dry against the skin of my neck, on the backs of my arms. My heart pulses steadily within its sheath. The sky is fat with promise.

Quickly, the pavement is again rattling by underfoot. Again, an odd sense of remorse has settled inside me. And again, I have Interstate 70 mostly to myself. I draw one breath, then another, and then one more. Twenty miles later I've nearly forgotten the owl. Instead of hooks and beaks and bones like flutes, my mind is filled up with old fears and budding uncertainties.

About five years ago something happened, something big. A star in Cassiopeia collapsed, or didn't, as it should have; a flower failed to bloom, or blossomed a string of passionate red petals; a stone, pushed by spring floods, left the spot where it had rested for 3,000 years, or didn't; a river leaped from its bed or failed to—something big. And at that moment, everything, I mean *everything,* changed. Five or so years ago that happened. I hardly noticed then. But I can't ignore the evidence any longer.

I've been a biologist most of my life—specifically, an immunologist, but I like the term biologist better. Maybe because it doesn't seem so closely linked to human medicine, maybe not. Maybe it just sounds earthier. But from about the time I was twelve, a biologist was what I wanted to be. I still want, sometimes desperately, to be a biologist, I'm just not certain anymore what that means.

And it's not that I'm leaving science, exactly. No, it's a whole lot more like science is leaving me—more like the universe one day just said, "That's enough of that shit." And biology, having never been particularly pleased with the job I was doing anyway, slipped out the side door.

Off the road to the right a golden eagle is hanging in a morning thermal, watching the land below—waiting, almost indifferently, for a single lapse, an instant of forgetfulness, a moment of courage. And the eagle knows—no matter how many times before everybody noticed him, no matter how many times before no one mistook him for a cloud or a raven or a swarm of bees—there will be mistakes, soon. The eagle waits, in the shaft of sun-warmed air, and circles.

This section of I–70—after you've left the cities and the river behind and climbed up onto the eastern edge of the Colorado Plateau—to the extent twin ribbons of concrete can, soothes me. The Manti La Sal, snowcapped this January, pierce the sky to the south, tattered old turbans still clinging to their heads this morning. The north is rimmed by the Book Cliffs, laid out here like some new attorney's collection of *The Law Review.* All else is sagebrush and sand punctured by an occasional juniper, but otherwise unmarked, in large part, by man or beast. And of course, there is sky—lots and lots of sky.

Biology taught me my first serious way of looking at things, a deep way. A little like the way the first older woman or man you sleep with shows you how something you've hardly even noticed for most of your life can give you unimagined pleasure when touched very softly, when squeezed a little too hard, or when brushed by human lips. And that's probably good. But after an experience like that it's easy to become obsessed and try to build walls around things, try to isolate a thing—tear it down, build it up, come to know it apart from every other thing in the universe. And then there comes that time in the middle of the night when you wonder, no matter how you try to suppress it, wonder—just for a instant—if it's enough; wonder if maybe you're cheating yourself somehow. Or at least I did. And that's all science was waiting for, that's when science tossed her ring on the bed and stepped out the door.

Now, I don't think she's coming back, and in the 3:00 A.M. darkness of winter mornings, that sometimes scares me.

The turnoff to Cisco interrupts my thinking. Good. I pull to the right, loop south under the freeway, then head west on a narrow little road that ultimately leads to Cisco, Utah. That's a lie, of course—all roads, like rivers, *ultimately* lead to the sea. But in this neck of the woods this road's primary purpose is to transport people from points east into the Cisco metroplex—in the middle of nothing but sagebrush and, for no apparent reason, a collection of ten or fifteen wooden buildings in various states of disrepair and habitation. Some are posted with sternly worded signs that read redly NO TRESPASSING! I can't imagine why anyone would trespass here anyway—in one of these ramshackle wooden buildings, likely full of kangaroo rats and garter snakes, black widows, scorpions, and bats—in the midst of more public land than you or I can even imagine, and every square meter of it just begging for a footfall. But Moab attracts a different kind of crowd these days, and much of that crowd, evidently too much, gets to Moab via Cisco. So who knows.

Before those signs went up I thought this would be a nearly ideal place to live. Middle of nowhere—no telephones, shopping malls, stoplights, school boards, or chambers of commerce. That may have changed. Seems unlikely these men and women would go to the trouble of putting up signs without cause.

A quick left and I'm on the road to Moab. The road from Cisco to Moab is a fine one, full of odd twists and turns, narrow spots, overhangs, washouts, abrupt ups, sudden downs, unexpected lefts, unanticipated rights, and lots of river. The powers that be will probably widen and straighten it any day now. For the moment, I just drive as fast as I can, honk my horn a lot, and imagine how it would be to end my life in this here river.

The road winds past some truly unbelievable places—Dewey—with its inspirational messages, Proudfoot Bend, the Fisher Towers, the Professor Valley. At the end, though, the road just settles into a slow waltz with river and winds through a deep trench of Entrada and Navajo sandstone—twin spirals of petrified earth, their dark red reflection bleeding into the green riffles of the Colorado River. Easier to see here how the river got its name. A wonderful spot. Then, all too abruptly, you're

The Professor Valley, on the road from Cisco to Moab, Utah

spit out onto State Highway 163, down from Crescent Junction. Sic transit gloria mundi.

I hang a right and pick up the road toward Crescent Junction. Only briefly though—a quick left and I'm onto the Potash Road. There's a little canyon up here I noticed on an earlier trip, and today I'm going back to see if I can find a way in—and out, of course. That's important too.

Past the Moab slough and beyond red-rock walls bristling with bolt ladders, about twenty minutes down the road I locate the spot I'm looking for—a little neck of stone, a bend in the road. I pull off to park. Rain and maybe snow were in the forecast I caught on the radio this morning. But this December day appears to have something else in mind. It's a bit overcast, true. But according to the little thermometer on my pack it's nearly fifty-five, and my feet are anxious for sandstone. I change into my boots, stow the keys under the seat, and fill my water bottle.

As a mixture of Cache la Poudre and Colorado River water trickles into my bottle, I can't help but think how glad I am to be here. The air smells of river and stone, of blue sky and wet willows. My skin feels like someone else's. And I just know something remarkable is about to happen. Not because of omens or auguries, though there've been plenty. I just know this place. Something is always happening.

Much of my time in science was spent trying to make something happen—animals get sick or better, data fit or not my preconceptions, nucleic acids rearrange or not. It isn't like that here. I never know what's coming, true. But I know for sure something is already in the pipeline. I just know it. I know, too, that it's for me and me alone, for certain, every time. I shudder—a victim's spasm of anticipation—shoulder my pack, and head out.

Canyon reflections—Entrada sandstone near Moab, Utah

The trip in goes up a little wash that quickly spirals into a cleft between two high walls of Entrada sandstone. The bottom of the cleft is filled with deep red sand. Delightful to look at, a pain in the ass to walk in. Walking in sand is almost like dancing—almost, that is, like waltzing, waltzing with someone you don't much care for to the music of a band that only knows tangos. As soon as I can I work my way up onto a shelf on the west wall and traverse past dry falls. The canyon widens then, opens out to a broad stone hogback littered with black rocks—about twice the size of my fist—pockets of sand, and a handful of mint-green junipers. I stop for water.

The sky's a cirrus custard, and there's a little breeze blowing up from the river. You can smell the water, and something else—something perhaps not as old as water, but near as old as these rocks. It is a fine day, I mean a *fine* goddamn day, for a walk.

Across from where I sit (two canyons away actually, a thing I will learn later) there's an enormous jug-handle arch. Hard to say from this distance, but the arch must be a couple hundred feet high and forty feet thick. The sort of handle that you might find on Thor's jug of mead or Godzilla's jug of fire-breath fuel. I think of water shaving away at that sandstone, trickling in in September, blowing its hydraulic jacks in Janu-

Jug Handle Arch, near Moab, Utah

ary, then pulling up stakes and slipping away in the springtime—year after year. One precious millimeter at a time. Water is a powerful mentor.

I shoulder my pack again and head off toward the rim of the near canyon. Between here and the stone ledges above the canyon there's a minefield of cryptogams, little mixtures of algae, fungi, mosses, lichens and what have you, that set the stony soil up for the entrance of bigger things—grasses, trees, and Howard Johnsons, for example. Without them, what little soil there is up here just blows away. Because it takes about a hundred years for these little guys to become sufficiently established to hold back a tablespoonful of soil, it seems . . . insensitive to crush them underfoot simply because they lie between me and my goal. So I try to avoid them. Walking in sand may be a little like a waltz to tango music, but walking across cryptobiotic soil is a lot more like trying to dance a samba to polka music. Hard to find a good rhythm. From a distance I imagine my gait must look like I've some sort of neurodegenerative disorder. Like there's something not wired quite right, or soldered securely, between spine and brain. Which is true, of course, but not the reason for my stride.

Finally, I'm out onto sandstone again. Here, the Entrada formation rolls over toward the canyon like the crust on a loaf of rock bread, then the sandstone gives up any pretense of horizontal and drops straight into the canyon below—fifty or a hundred feet. I drop my pack. When I was here before I never could find a route into this canyon that looked any better than the one in front of me.

I rig a rappel anchor, then uncoil my rope and toss it into the canyon. I clip the rope to the anchor, pull on my harness, and work my way down the vertical stone wall into the canyon.

Rappelling always makes me mildly sick to my stomach. No matter what my mind is doing, my ass knows that it should not be in the lead when together we are about to step off a cliff. My ass knows this is wrong, deeply wrong, and it reminds me of that. But my mind is attentively occupied elsewhere—noticing each of the grains in the sandstone, its size and texture; noticing the scrub juniper, seemingly growing out of solid rock; feeling each of the movements of my legs, the pressure of the harness, the heat of the rope slipping through my hands, my anal sphincter, the grate of my boots against the rock, the wind across my face; and underneath it all, the odd sense that I might—for no reason at all—just let go. But once again, I don't.

And then I'm down.

Maybe awareness is the answer, the answer to the riddle of life, that is. Untarnished awareness.

Because at this point I'm not sure there's another way out of this canyon, I leave the rope in place, stow my gear, and start up-canyon. This canyon is remarkably water-rich, especially for December—willows and tamarisk, reeds and junipers, and a thin stream of windowpane-clear water grinding at the stones and curling its way to the sea. Again I feel the Artist's brush along my arms and up my spine. A fine day, a truly fine day.

As it rises, the canyon cuts more deeply into the stone, now five hundred, maybe a thousand feet below the sky-capped rim. Huge red slabs, streaked with black, rise like Lazarus, spattered with water and maidenhair fern, dry and gray this winter.

A raven screams somewhere.

I cross the stream for the last time, and I notice two sets of prints in the sandy soil—one deer, one human. So there is another way into the canyon. The human might have come in as I did, but I doubt that person carried the deer in on his or her back. So I suspect the bottom end of this canyon can be breached. And though both sets of tracks are old, they annoy me. I mean, after all the trouble I've gone to, I still have to share this canyon with the image of another human being. The clouds thicken a little, the wind blows a little colder, and my left foot begins to throb slightly in my boot as I work my way through the undergrowth toward the upper end of the canyon. Why can't other people leave well enough alone, dammit? I think of turning back.

But only a mile or so farther the canyon comes to a dead end a thousand feet below the earth's surface in an alcove of vertical stone. There's a pool, and more maidenhair fern—but unlike those lower down, many of the ferns here are green, hard into December. It's been months since I've seen anything but browns and grays at home. This verdant green rolls around inside my eyes like pearls on an oyster's tongue, and inside my eyes this color of a faraway spring stirs something deep down and very old.

I unshoulder my pack and sit. A long sigh slips from my mouth. Above me the stone rises thirty feet, underhung, thick with ferns, then ledges off, back to the real wall. There the sandstone spires vertically to a frosting of dark caprock, and then to the sky, a sky you can hardly even imagine from so far below. And at the center of it all the pool, black-green, rippled by the wind—deep water that tugs at my eyes and strains my imagination.

I wish Gina were here. She has been as long as I without green. But it is different for her. At times, for her, being without green is like being without air, without sight, without the touch of a human hand. To bridge the brown ditch between autumn and spring, she fills the sills of our winter kitchen with colored bottles sprouting paper-whites and clay crocks with amaryllis rising to crimson spears. And that sustains her, but only just, until we plant pepper and tomato seeds in February and fill the bedrooms with fresh green sprouts. She would like these December ferns, she would like them in her eyes.

Knowing things. Knowing things that no one else has ever known. That's what science was for me. Even a little thing—the location of a bit of sugar along the spiraled shaft of an important protein—known only to you sets you apart, apart from all who came before you, apart from all who surround you. Knowledge, and the power to create it. That's a calling. Hell, that's a sacrament.

But what about the rest of it? What about the power to see what all the others have seen, and to know in that instant what they knew? What about the power to remember—to hold in your cup knowledge deep as these rocks, and to recall who we have been? What about that?

Overhead, the sky has cleared and sunlight is seeping through. Shafts of light work the ripples in the pond. Bits of blue appear in the shredded fabric above. I wonder how long this pond has lain here. A million years, 10 million years? I wonder how these ferns can be this green, deep into this December.

As I sit, the sun on my back, the water at my feet, a snake of contentment uncoils inside of me—an odd sense that, for this moment, it doesn't matter, any of it. I lie there, an hour, maybe an hour and a half, listening to the clouds slide past, to the wind in the pond, the steady seep of dark water through slits and notches in the ancient stone. Then it slithers out again, the snake—back into the red soil, the malachite-green water, the sunbaked, brick-colored walls. I reach for it, but it's gone.

I stretch and rise to leave.

I think then to give something to the pond. Something in return. But I've left the owl's feathers in the truck, and I have nothing else today, nothing a pond would want anyway, nothing of mine that would interest old water.

I think of the stones in the bag hanging on my pack. I pull out a pebble of basalt, a smooth black fingernail-sized chunk I've carried for a long time, and I toss it into the pond. There's a satisfying sploop of water, and then the pebble is gone. Ripples move out in widening arcs until they strike the pond's banks, reflect back on themselves, and slowly die. I start down-canyon.

On the way in I noticed no other exit from this canyon. And though, as I said, there must be one, I head instead for my rope. I intend to drive to Blanding this afternoon, and the light is beginning to fade. I find my rope where I left it, a thing that always relieves me. You wouldn't think anyone would move a rope left like this one, but people do the damnedest things.

However, because I hadn't anticipated using it to get back out of here, there's another dance I have to do to "rescue" myself. I tie the bottom of one end of the rope to a stout little juniper. Then I slip into my harness and tie on my ascender. I clip onto the rope, work my way up the cliff face, over the edge, and onto the top. There, I tie a quick figure-eight knot into the doubled rope at the top, clip it back into the rappel anchor, slip the doubled rope into my descender, and rappel down again. At the bottom for the third time, I untie the end of the rope from the juniper, clip into my ascender, and climb the sandstone face once more. Saved, I sit for a drink of water and a breath or two.

But I don't really think the secret of life is locked up somewhere inside of knowing something no one else has ever known, or even inside the sacred crate of common knowledge, not really. Maybe it's buried instead among the things, like this rope, that tie us to all the rest of it. Things that bind it all into a single thing—one fact, a solitary process, a lone event—the sagebrush beside me, the juniper that held me, the sky that warmed me, the stone I'm sitting on, the view just beyond the next hill, and the fibers that run between them all.

And maybe that ain't it either.

But I don't think science offers any deeper insights than poetry. And that's probably why science packed her bags. Equivocation can sometimes lessen a lady's enthusiasm for an obsessive-compulsive lover like me.

Regardless, it's a hell of day for a walk. I coil and stow my rope, my harness, the hardware. I shoulder my pack and start back toward the lower canyon. As I'm walking, for no reason I can imagine I recall a baseball game—St. Olaf's versus St. Patrick's, Catholic Youth Organization league championship. I struck out twelve batters that day. We still lost, of course. But I was brilliant!

The breeze is at my back now.

The Dog at Christ's Feet

Jorgé made pictures using wooden matchsticks,
Pictures of the Virgin with her Child, pictures
Of the Savior with his Dog. It was really Jorgé's

Dog, of course. You could tell that just by looking. But
Jorgé knew for certain, Jesus would have loved that dog. So
Jorgé put him there, at Christ's feet, and arranged Christ's

Right hand so it held the dog's head. The
Dog had its eyes turned toward Christ's eyes. Its
Tongue was hanging from its mouth. Jorgé

Taught himself how to do that while serving five years
Of an eight-year sentence for moving seventy kilos of cocaine
From Cheyenne to Denver. Each matchstick was glued just

Right to the foil backing. So that one stick would look just like
The firm edge of the Virgin's middle finger, another would be slightly
Curved just like the crease in the chin of The infant of Prague, or

Splintered so it looked just like the thorns in Jesus' crown
Or the holes in Jesus' hands. All of the matchsticks
Were burnt, charred on one end, so Jorgé's pictures wouldn't

Unexpectedly explode. Taped to the backs of the pictures were
Poems. Jorgé hadn't written the poems, though. He'd only thought
Of them. Jorgé couldn't write. Ray, Jorgé's friend, had typed

The poems for Jorgé on the prison typewriter. That typewriter
Was given to the warden by a retired schoolteacher who lived in
Florence. She thought felons should be rehabilitated. Sometimes,

The warden agreed with her. But that wasn't the real
Reason she'd donated the typewriter. She'd given the warden
The old Underwood because it had a problem. It typed "o's" that

Looked like "e's." So when Jorgé meant for his poems to say things
Like: I remember our home washed in moonlight, or: our love
Is like a downy dove, the typewriter wrote—I remember eur

Heme washed in meenlight—And—eur leve was like
A dewny deve. Ray didn't care, they weren't his poems. Jorgé
Didn't mind either. He couldn't read. Jorgé just liked the way

The words looked walking across the backs of his pictures. Jorgé
Gave the pictures and the poems to his mother and a woman
He often thought of late nights in his cell. Jorgé's mother Lupe

And his girlfriend Rosa liked Jorgé's pictures okay. But Jorgé's
Poetry scared them. Their letters didn't say that though. Their
Letters said they loved the pictures, and they would come

To visit Jorgé, soon as they had the bus fare. Jorgé
Wrote back: I am leeking ferward a let te seeing yeu seen. Rosa
And Lupe stayed at home. Jorgé went on, though,

Looking forward and late nights in the blackness of his
Cell popping stick matches with his fingernails and sculpting
The pictures he saw inside the flares. And Ray went on too,

Went on typing out the words that Jorgé said were buried
In the little wooden sticks. The old Underwood went on as well,
Putting Jorgé's words on heavy sheets of prison paper.

The Savior's Dog, though, took to keeping its tongue
Inside its mouth. Then moved its head from the Savior's hand,
And finally disappeared altogether from in front of Christ's

Feet in the foreground of Jorgé's matchstick pictures.
That puzzled Jorgé some. But Ray didn't mind.
They weren't his pictures. It wasn't his dog.

All goes onward and outward, nothing collapses,
And to die is different from what anyone supposed, and luckier.
—Walt Whitman

In the End Death Looks a Lot Like Me

My body knows there is water beyond the low hills ahead of me, lots of it. My body knows that, because it holds a river in its chest. But my mind sees no sign of water, only rock wrinkled up like old squash, only sunlight the color of blindness. If I were thirsty and my water bottles were dry, my mind might leave me here to die or, worse, send me after snow at the 10,000-foot summit of Mount Hillers in the Henry Mountains to the west. A mind is a terrible burden at times.

It is already warm this morning, though it is only April, and as I lace my boots a bit of fear wriggles inside me. I don't know where it comes from, this dis-ease with the unknown, this nugget of angst. But it always shows up to see me off.

I look for my keys, check the spare set behind the bumper, look to see where my wallet is, check maps, compass, water, rain gear. Out of excuses, I start toward the canyon, images of the 20 million acre-feet or so of Lake Powell—the water I know to be at the other end of this canyon—and the laughter of the long red ribbon of the Colorado River twisting inside my head.

About a hundred yards downslope from my truck I find a way through the smooth rock wall—Navajo sandstone here, I think—a 200-million-year-old relic of the vast Navajo desert that blanketed this region back when trees were ferns and people were cartilaginous fishes. Sand become stone under a 100-million-year mud pack. Stone become canyon as Laurasia disintegrated and the continent warped up like bad cheese.

Then, this place might have been green and swampy—splattered with bogs full of lily pads—then, this might have been the sort of place frogs would like, a place where frogs would thrive. Then, but not now. Surely not now.

The sun-baked rock stares back at me sullenly. The canyon deepens quickly, sixty, maybe seventy-five feet, and the Navajo formation has given way to the butterscotch Kayenta, and now it yields to blood-red Wingate sandstone. Below the surface here the air is warm and still, and as I walk I imagine I hear the lake lapping against the Chinle bench miles below. I imagine, too, the slow drip of snowmelt in the Never Summer and Wind River Mountains and the water, far off, chewing steadily at the concrete and steel ribs of Glen Canyon Dam.

Canyon hiking is a wondrous thing. I used to be a mountaineer (well, sort of). I spent most of my hiking time in the mountains anyway. But four five years ago, change smoldered inside me. First I noticed how the bottoms of most desert canyons are slightly less steep than most mountainsides—a more than trivial issue for a man committed to too little exercise and too much beer. Then I began notice the ways the desert has of showing you just how little it takes—eight or nine feet of polished stone, up or down; too little water; quicksand; spiders, snakes, and scorpions; thunderstorms; rivers of mud . . .; how little to turn an hour hike into a day hike, or a day hike into a final hike. Intriguing. And in the end, I found canyons and rivers that defied imagination, grottos of ferns and monkey flowers, riffles and warps written into the ossified breast of the earth, and the intricate ways of desert stone.

I spend less time in the mountains now. (And no, it isn't because of the time with Ed when I scared the bejesus out of myself 100 or so feet up the 4,500-foot North Face of Edith Cavell, or the time on lead I set every chalkstone on my sling to make a rappel anchor in the middle of a route my grandmother could have led, and then fought for long minutes to control the panic so my friends wouldn't have to pry me from the rock with an ice axe.) As I said, it's just that mountains and I are on slightly different terms now, and the desert has given me glimpses beneath her ruffled skirts that stirred something deep inside this old body.

And today, a Wednesday I think, there is a canyon wren. Canyon wrens are a thing I could have overlooked if I were the only god. I would have thought of making these fine long canyons, certainly. But I might have forgotten to add the song of these wrens, and that would have been a very bad thing. The air fills with music and the gingerlike smell of roasted rock.

And although it's true that I rarely find what I imagined I wanted when I set out into the desert, I rarely return without something I needed—a stone, a bird's song, a thought, a view deep into the meat of the earth, a single red blossom amid a sea of yellow stone. Carrying sometimes gallons of water seems a small price to pay. Besides, I take shorter hikes now, and I'm getting better at finding old water.

The canyon has widened some, the bottom is now fifteen or twenty yards across and strewn with condominium-sized chunks of the Wingate. Junipers, ripe with berries, rise dark green against the crimson stone—the trees, the stone encouraging each other.

Deep inside the Colorado Plateau

About a mile farther I stop to tighten my right boot. As I'm relacing I notice, no more than a yard beyond my boot tip, a frog—a desert tree frog. This part of the canyon is dry, and has been for a long time. I have no idea where this frog came from, but he's moving very slowly, very far from water. I prod the frog with the tip of my forefinger. The frog moves sluggishly a few inches, then stops and stares at me froglike—no fear, no affection, only bottomless indifference. The frog is, at most, a few hours from death—there is no water here, no mud.

I consider for a moment taking the frog with me. But I have no idea how soon I might find water, or whether I would only be taking the frog to die in my shirt pocket instead of among the rocks. Furthermore, our acts are not without consequences—as the raven circling overhead knows full well—and I do not relish responsibility for another being this day. I leave the frog as I found him, desiccating slowly, and move off down-canyon, promising that I will move the frog to water if he is still there on my return. I focus briefly on images of the distant river. Billion-year-old gravel grinds beneath my feet, and my thoughts clear some.

The narrowing gorge turns cool and shadowy; now the base of the canyon is peppered with chokestones and there is water in the low spots. I will get the frog on the way back.

At each of the larger stones lodged in the bottom of the canyon I evaluate carefully, lower myself, and then quickly climb back the way I've come. Assured that I can return, I climb down again and move on (somehow I'm comforted by this). At last I come to a large chokestone, a Peterbilt-sized piece of rock the color of old blood, lodged square in the throat of the canyon. Beyond and below this stone there is a pool of dark water. From top to bottom perhaps eight feet, maybe nine, of bottle-smooth stone, then three or four feet of water. Beyond, a stretch of deep narrows fills quickly with shadows and vanishes into the rock. I sit to eat.

By my calculations I can't be more than a mile or two from the lake. I would like to see the lake. Worse, there are a pair of boot tracks emerging from the pool below and sauntering off, down-canyon. I measure the distance to the pool by eye once, twice, then again. I consider jumping. But then I see myself standing in the water below the stone, my hands raised Joshua tree–like, I see myself slowly starving, hoping for a flash flood to silence my pathetic whimpering and end my aching hunger.

I walk back to see if I can find a route around this stone and pothole. About twenty yards up-canyon I find a way onto a ledge above and follow the sloping sandstone until I am nearly over the chokestone. Here the ledge tilts off at sixty, maybe seventy degrees—an easy walk, if you are six feet off the ground, less so at thirty feet, less so alone. Another vision of me—lying on the canyon floor with a broken leg and urine-stained shorts, ravens plucking at my eyeballs—rises behind my sweat-beaded brow. I look for a long while at the route across the rounded stone. I step forward. I step back. I imagine the satisfaction of reaching the lake. Then the eyeless vision reappears. Forward, back. Finally I give up and turn toward home.

Author approaching end of line near Hanksville, Utah

At the foot of the wall I meet an old friend. One who waits for me in places like this, waits to remind me of what a cowardly asshole I am. I've collected a lot of justifications to counter his arguments, but he always wins and leaves me in a shroud of guilt about giving up, no matter the reason. I promise myself I will come back one day and finish this hike. My friend scoffs. I begin working my way back toward my truck. He walks behind, nagging.

On and off as I walk, I look for the frog I left behind. Another wren, or maybe the same one, follows for a while, turning a warm breeze to poetry. As always, the trip up-canyon seems longer than the one down. Finally, I pass a point that I'm sure is above where I left the frog. I abandon thoughts of making up for leaving him behind.

I can't help but wonder, though, what became of him. At the end, was it a white light, or a sudden cloying darkness? Were there regrets? Did he, at the end, wish things to be otherwise? Was he afraid? And finally, had he imagined, this morning, that death would come for him today, and that when it came, death would be dressed like a middle-aged man in old leather boots and frayed walking shorts?

I nearly step on a snake.

The snake moves off a respectable distance and then stares at me as though appalled by my oafishness. Ever since snakes discovered that sidewinding thing, they've been less than tolerant of most other forms of locomotion. It's a small gopher snake, not nearly large enough to have ended the frog's thirst. But the snake reminds me again of circuits beyond my comprehension.

As though it mattered, I apologize—to the snake, to the frog, to the guy nagging at me from behind, and I walk on.

Near the end, as I work my way out of the canyon, I notice lion tracks in the thin red crust left on the soil by the last rain. The prints are large, large as my hand. How did I miss them on my way in? The saucer-sized indentations lead directly to my truck. Was the cat here while I was away? The tracks seem too old for that, but why didn't I see them before? I imagine the cat at my truck, sniffing at the tires, looking for some sign that I remembered, that my sweat or my urine remembered. I imagine too the cat's hunger.

I stop and look at the old and rotten country that surrounds me. I think for a little of when this land was young, the time when people were bears and wolves, the years when men walked on four legs and counted fear among their friends. I think of volcanoes and the awful sound of the stars at night.

I scan the hills nearby. No sign of the cat. Nothing. Nothing at all but a bone china sky and sweet empty air. I dump my pack into the back of my pickup. Aloud, I retract each of my earlier apologies. Then I pull a cold beer from the ice chest and settle, a little arrogantly, onto the tailgate of my truck.

Natural Selection

People, like you and me,
weren't always people.

No,
long ago, people
were trilobites and horseshoe crabs.
Then, we were fishes. Trees

were mostly ferns then.
And later, people were crocodiles
and scale-plated lizards.

Then after the fire, we were
    birds. . . . This
was a wonderful time for people.

But, my favorite time
was much later, the time
    when people were bears.

    The way we would roll
through the woods on strong legs,
eating blackberries

and mice and deer, eating
roots and elk and rabbits. The way
we'd shit wherever we wished.

Oh, and the way we fished,
    for hours, with our bare hands
in the stinging cold water, then the taste

of that cool pink flesh. How we'd
sit up all night
singing the old bear songs.

Then just before dawn, we'd set to work
making new bears. A few of the elder
bears could still do this
using only the warm clay
of the forest, but those of us,
too young for these ways,

made new bears from our own flesh
and the urgency of our couplings.
    Our cubs had the sharpest teeth

and the sweetest breath. . . .
    But that was very long ago.
That was when bears could speak.

Few people
are bears now.
Most of the grizzlies

and the blacks you see
are only bears
now.

And people?
Well, mostly
    we're just people now.

"Because I am mad about women
I am mad about the hills."
Said that wild old wicked man
Who travels where God wills.
—W. B. Yeats

Because I Am Mad About Women

There is something about pickup trucks and dirt roads—some of each in the other, I guess. And this morning the bergs of slickrock in the red sand only tighten the bond, only narrow the distance between truck and path. A moment, it seems, when the earth aspires to the freedom of the mechanical and the mechanical longs for the simplicity of the earth. And above it all rise the hard blacks and browns and ochres of the high Utah desert—a broken place, and because it is broken, a place where anything is possible.

Only a few miles from the hard road, a few miles of molar-loosening stone and tire-sucking sand, and the road spins itself out in the shattered waste of White Canyon—a godforsaken spot in the heart of the Colorado Plateau, near nowhere. The truck clicks and sizzles as it cools. I sit and draw sweet deep drafts of the blue Utah air.

I was near here last May, but the approaches I chose that day led to nothing but rotten ledges of buff-colored stone hundreds of dead-vertical feet above the riverbed. Repeatedly rimmed-out and finally pissed off, I spent the short end of the morning heaving boulders and bits of old trees into White Canyon. The canyon ignored me. So after a while I quit and drove off to Hite to see what was left of Glen Canyon and the Dirty Devil Gorge, to look in on the near edge of Dark Canyon.

Maybe five miles west of where I'd spent the morning rim waltzing and boulder herding, I passed two or three roads that clearly coiled all the way to the bottom of White Canyon. Even from the highway you could see that. Though I'd tried to insult it, to bruise it, and finally to ignore it, the canyon had shown me another way—kind of Zen, huh?

I made a note of these roads on an old map as I drove, nearly hitting some sort of large rodent or small dog. And I revised the itinerary for my next trip to southeastern Utah while swerving off the ratty shoulder and back onto the pavement, sending a rooster tail of Utah soil into the noontime sky. I should learn to stop while I write, but hell, I don't even stop while I read.

When I actually follow through with plans to return to a place I've been before, it sometimes amazes me. So often, I become infatuated with the ankles and thighs of some new place, and the old spots turn shabby and fall from my mind. But here I am, on this glass shard of a July morning, staring down the throat of White Canyon from the donut-and-coffee-stained seat of my truck.

Overhead, someone has unfurled one of those royal blue tapestries woven only in the rich air above the Colorado Plateau. No cloud, not even a vapor trail to mar the perfect dome, and somewhere nearby a canyon wren is rehearsing the song it will sing on the final morning. God, it is good to be here!

White Canyon, near Hite, Utah

But, it *is* July. So I fire up the truck, push it into low-range, and wallow through the sand and gravel until the truck is three or four feet above the riverbed. Not really enough to make any difference in a genuine flash flood, but high enough to relax my mind this morning. The truck seems comforted as well.

White Canyon lies like a snake along Utah Highway 95, its head among the stone spans of Natural Bridges National Monument, its tail twitching in the dark pools of Lake Powell. Along the way it gathers water from canyons rarely visited by people, canyons with names like Cheesebox and Hideout. And along the way, White Canyon cuts through 60 million years' worth of sandstone and a few hundred feet of elevation. At times, times when the sky wrinkles overhead and the winds wail, this canyon gathers water like a fist and downstream from here carves some of the darkest, dankest narrows on the Colorado Plateau.

One of these days, with the aid of the Lord and an old rubber tube, I aim to swim those narrows, but today isn't that day. Today I am headed east, up this canyon, into the narrows of Fry Canyon and maybe the bottom end of Cheesebox.

The day just keeps smiling back at me, but I can't shake a feeling that I was here before. No sense of when, but clearly not recently, and for some darker purpose it seems. . . . I can't flesh the image, so I stop prodding at the old cerebellum and gather my gear. Oddly less anxious than usual, I start up the canyon.

My sense of history with these canyons is no doubt sharpened by time spent yesterday among Anasazi ruins in Mule Canyon—typical stone and mortar huts wedged beneath great overhangs of Cedar Mesa sandstone. Mostly empty for the last seven hundred years, but in the corners you can still find cool shadows, old smells, odd sounds like words whispered between stones. Dry now, very dry, the people long gone, and no one knows why, no one alive anyhow.

Some of the granaries and dwellings hold only deer-mouse droppings or flakes of clay and shale, others are paved with layers of dust scraped by the wind from the rocks overhead and the riverbed below. The windows are small, most too small for head or shoulder, always with views toward the canyon, toward coyote and lion, and all the other dangers, real, imagined, or both.

The dances though, and the harvest, the hunt, those are the times I would have enjoyed most, and the nights, nights when the sky burned with the awful light of Orion, and the stone hut filled with the warmth a dying fire and the moisture of a soft-breasted woman close beneath coarsely woven blankets. . . .

White Canyon takes a sudden turn to the left, and my attention is snared by a horned owl's feather twitching where the wind pushes between the rocks ahead of me. The air has warmed, the canyon deepened. I retrieve the feather from its gravel bed, stick it under my hatband, and stop for water—tepid already in the July heat. I sit staring into the fire of the gorge.

There is a thing I cannot name about canyons, a feminine thing. You feel it most strongly in shadowed places wet with old water, places where night air still lies cool above tadpoles and water striders, beetles and curled green moss—fertile, feminine.

The wind won't leave the owl's feather alone. Like a child, it pulls and twists at the shaft. And the feather, as though it remembers, clutches at the air, reaches for the sky above the canyon wall. I pull it from beneath my hatband and slip it into the windless top of my pack. I put my water bottle back into its place as well, shoulder the pack, and start back up-canyon.

The canyon alternately narrows and widens, steep walls of chocolate-colored or ice white Cedar Mesa sandstone capped with Organ Rock shale. Mired in the Permian here. But far above, occasional glimpses of gray Shinarump and iridescent green Chinle formations laid down only 200 million years ago, Triassic—thunder lizards, birdless skies, fern-and-frog-filled forests. Shinarump is a combination of old Indian and perhaps English words that translates roughly as "wolf-butt." Good name. In places, I think this stone looks quite a bit like the south end of a northbound wolf, in other places less so. Stone is like that, irresponsible.

To burden my mind, I look for routes that I might follow out of the canyon if my disease, or a distant thunderstorm, got the best of me. There is none. Every apparent route up the canyon wall terminates in a vertical smear of sandstone or a rotten ledge of shale, undercut. No way out. I must go where the canyon goes. I stifle the wisp of panic that curls from an ancient and evil wrinkle in my mesencephalon.

Perhaps it's the sun and the constant pounding of my feet against the rocks. My thoughts seem to arrive from somewhere else today. Again I stop for water, this time in the shade of a massive overhanging wall maybe four feet above the sand of the river bend. As I sit beneath its bulk, I imagine its crushing weight in a sudden collapse—quick, dry, a good death. But what if you weren't quite dead under all that stone, legs broken, fingers snapped, but still breathing from a space no larger than your fist? What about that, huh? How would death come then? Shock? Dehydration? A suffocating panic?

The stone, beneath its layers of brown and black varnish, remains in place. I empty the first of my water bottles. Cause for contemplation. I've really no idea of my destination today, and I've seen no water in the canyon. I take a long warm pull from the second bottle and watch a lizard, one of those gray ones (*Grayus unoensis*), skitter out onto the sunlit face of the stone, stop for calisthenics, and then move on, out of my sight. I shoulder my pack again and follow the lizard into the sunlight, into the canyon. Ninety degrees according to the little thermometer on the zipper of my pack.

Nearly four hours since I started. Seems I should be somewhere near Fry Canyon by now. I imagine it around each bend in the canyon, at least at each bend to the left, because I know the canyon enters from the south. That's the way I hike. In some of the most remarkable country short of the dark side of the moon or the poles of Venus I focus only on what lies ahead, what lurks beyond my sight.

I stop, force myself to look around, carefully, breathe deeply the stone-dry air. I pick a pebble from the riverbed and put it in my pocket for company. Always better walking with a friend. Up ahead a raven calls. I begin again, more slowly, more attentive.

My mind wanders from the canyon to a reef of odd, heat-driven daydreams. I wonder at the skills of the hawk and the raven. And I think of afternoons I spent listening to the small roar of white-throated swifts as they carelessly spiraled like little buzz bombs into a thousand or more feet of empty air over Soda Basin.

One of two poor trades made along the evolutionary trail—feathers for big brains, that is. The other, of course, was the os penis for a vertical spine.

I stop, again in a place shaded since morning. In the relative cool I drink the last of my water. Briefly I consider turning back, but I don't imagine I've come so far I couldn't walk back without water. Imagination is like that, full of shit. After a bit I continue waterless up the canyon—a little more motivated now, a little more goal-oriented, a little more like Randall Patrick McMurphy after his chat with Nurse Ratchett. The stones seem larger, sharper, hotter.

An hour or so later a side canyon finally breaks from the main gorge and slices off to the right. It must be Fry, no others on the map. I stop to admire the change in the riverbed where Fry Canyon dumps its litter into White Canyon. Signs of humans here. Some old plastic sheets, a couple of rusted cans. I wonder if the people at Fry Canyon Lodge (a place bearing a disturbing resemblance to the Bates Motel) are behind this, or if others stopped along the highway to the south to dump their trash. Immaterial and irrelevant, I declare—trash is trash.

Regardless, Fry Canyon opens alluringly. I turn and work my way beyond the boulders at the canyon's mouth, and immediately Fry Canyon narrows. Shadowy and cool, but only briefly. Quickly the canyon widens and warms again. Disappointed, I walk on. I had imagined this first tightening of the canyon was the narrows I was looking for. It was not, and the sun is again stroking the back of my neck, reminding me of the empty water bottles in my pack. Again, I think of going back. But not now, I think, not when I imagine I'm so close.

A half mile or so farther, the canyon again constricts. Two ravens suddenly fall from an unseen rest above me and scream loudly. The first turns and flies down-canyon, the other glides cross-canyon and knifes into the narrow throat of the gorge ahead. It screams again, slams upward, and is gone. I follow.

Here the canyon tightens wonderfully and twists off in sensuous curves of sandstone carved like rock ice or yellow-brown sea foam. I move into the shade of the narrowing walls, the temperature drops noticeably, and there is the taste and musky smell of old water—warm cumin and summer savory—in the air.

No sign of the raven. She must have spiraled out the narrow slot above, now dripping with sunlight. Farther in, the canyon squeezes even closer, vises down, darkens, four, maybe five feet wide now. The walls lurch up forty or fifty feet toward upper Fry Canyon. A final twist, walls within walls, and I come to a deep pool of dark green water lit by a shaft of light fallen from a circular slice in the stone above.

I drop my pack. My thermometer says it's twenty degrees cooler here than in the main canyon. Frogs, in all stages of frogness, twitch and jerk in the small pool. Water striders and beetles twist among them. There is no moss or fern, though—the deepness of the slot, the lack of sun, leaves the pool bare. Here, in the deepest recess, only wrinkled sand, the color of young skin.

I draw the water up in my hands and rinse my face and neck, my lips and ears. Water trickles down my chest and back, into my pants. The sun is forgotten. I soak my head, then fill both my bottles with water filtered from the pool. I drink long and deep, fill the bottles again and stow them in my pack.

Fry Canyon, near Fry Canyon, Utah

Last May I stood above this place. Overhead, Fry Canyon forms a deep V here and narrows to a polished stone slit no more that a foot or two wide. No access to this pool, this basilica, from up there, only vertical rock, slippery from a million years of running water.

I sit in the lush shade at the bottom of the canyon narrows filled with its feral water and think of nothing. For a long while, think of nothing. Then I take off my clothes and ease myself into the water. Waist-deep I sit. In the cold clear water, the pale skin of my upper thighs and abdomen wrinkles, my penis shrivels.

And then I think of other places. I think, for reasons I cannot touch, of my wife and my birth, of my children and the woman who bore those children, the smells and sounds of lovemaking. Beneath the thin green sheath of water I stiffen reflexively. All desire to move leaves. I sit for an unknowable amount of time.

Then thoughts of places farther up White Canyon begin to move in my mind, places I would yet see. One last thing, though. I reach for my pack, unzip the top, and pull the horned owl's feather out. Then I lay the feather in the shadows at the pool's edge.

Reluctantly, I gather my clothes and dress. I rinse my feet and pull on socks and boots over wrinkled toes and begin the walk out. Fold upon fold of wonderfully worked stone frames the fallen sunlight as I pass. Layers within layers, unfolded by wind and water and the tireless finger of time.

Then, abruptly, it's gone. The sun, the heat are again upon me. And a sense of loss rises like smoke from a freshly lit cigarette, then passes.

High on the wall of the canyon opposite where I emerge is a wonderfully worked collection of stone huts and granaries, carefully squeezed between two deeply curved outcroppings of stone. Again, the wisp of déjà vu. Then that too dissipates, and I walk on.

The Whiteness of Birds

Once, in San Jose del Cabo, where the last
of California crawls into a lapis-blue sea,
I came upon a shrine, it seemed.
Beneath a grove of wrinkled palms, a tree,
its knuckled branches hung with carcasses
like feathered lanterns,

carcasses of gulls, frigates, pelicans.
Most only bones, the intricate ironwork
of the great birds strung together with bits of cartilage,
hollowed fingers and vertebrae stretched
toward earth—fenestrated skulls,
delicate sockets.

And beside the tree, wedged beneath the low foliage,
a canvas shelter with once-bright,
now soiled and worn olive blankets,
a brown steel chair, and plastic bottles
filled with rust-colored water.
The dwelling

of the creator, I supposed, the supplicant.
On each of three days,
I passed before this tree, and every day,
to fill the void that flared inside my thoughts, I imagined
the things that flayed this man,
that clawed him where he sat—

the careless dance of birds
above waters too-filled with fish,
the insouciant whisper of the idle sea,
recollections of swollen women
beneath the icicle stars and the dry palm
thatch of warm palapas, . . .

But I think it was the birds
finally, the whiteness
of the birds themselves above a seamless blue world,
that brought him here, brought him
to a place where he needed
to be near their death,

forced him to gather the broken bodies
from the beaches, to sleep
beneath their flightless ruin, and there,
bolstered by their rot,
imagine an end to his shame.
I never found this man,

the worshiper of pelicans.
And the trees, the palms,
the broad-leafed bananas,
speak only of wind and salt.
But, as I sit now in his steel chair,
and watch the sea

breaking across the open land,
see the birds wheeling dizzily over schooled fish,
I know it isn't the bad in this world,
which is merely terrible, that distracts us—
in the end, it is the good, the beautiful,
that destroys us, for it is nearly perfect.

We need dreams the shape of lakes,
with mornings in them thick as fish.
—Naomi Shihab Nye

Cheese and Raisins

It was my mother who first told me that masturbation makes boys into homosexuals, that doorways draw lightning, that looking through someone else's glasses will ruin your eyes, that some people's fingernails are poisonous, and that only a thunderclap will make a snapping turtle let go of your finger or your toe or your lip. My mother was chock-full of information like that. But, wise as my mother was, Jennifer, my daughter, is the naturalist of the family.

At the moment, beneath an overhanging wall of Cedar Mesa sandstone large and portentous as tomorrow, Jennifer is sifting soil and shale back and forth through her thin hands. Abruptly, she turns up another bit of something and brings it to me for my consideration. The wall bellies out twenty feet above our heads like a gravid lion, then spires up three hundred or four hundred feet, indifferent tan with wide swatches of desert varnish spelling out words only rocks can pronounce.

"It looks like a vertebra, don't you think, or maybe a little bit of clavicle?" Jennifer asks.

"It looks mostly like a rock, I think. Maybe it's a rock bone," I respond.

"It's going to be like that, is it?"

She moves back to her digging, carefully working the earth with a pointed piece of juniper. She stops to move an ant to one side, saving it from a premature burial, pushes aside a couple of rocks, and then begins digging again.

Jennifer's squatting between two Anasazi kivas just now, completely absorbed by all there is to see. It took us four different shots at this canyon before we found a way in. And it took much longer to get here than I thought it would. In fact, it took twenty-seven years. But it was worth it.

The author's daughter, Jennifer Callahan, near Blanding, Utah

"Look at these ants here," she says. "I've never seen ants this small. Have you? What kind of ants do you suppose these are?"

"Piss ants," I say.

"Alright, I'll just quit asking." (She won't.)

I'm not like most of the women I've known. I can't put an exact date on many of the most important events of my life. I don't remember exactly when our dogs were born, or the day of the month when they died. I don't remember the date my wife, Gina, and I met, or the day of the week we became engaged. And I don't know when Jennifer was conceived, but her mother, Nancy, always claimed she knew.

I remember the pregnancy though. Jennifer's mother swollen and mysterious—her belly shaped and colored like a copper roasting pan, her breasts like brown fruit with nipples the color of morels after rain. And I remember the way Jennifer interfered with our lovemaking, so that I often ended up facing her mother's back while we were joined and cradling, with my right arm, the space where Jennifer lay. I remember too Jennifer's birth—all water and blood, and then Jennifer slippery and dark as an otter, crying only a little, and filling her eyes as hungrily as she filled her lungs.

"How can corn cobs survive more than seven hundred years?" And then answering her own question: "You know, though, they mostly dried the corn while it was still on the cob, and later stripped the kernels and ground them in metates for corn flour and meal. I'll bet that's what did it, that, and the just plain dryness of this place. It could desiccate the Nile. Look at this, it looks like it was piece of a basket!"

Jennifer and I, we've had our easy times and we've had our hard times. But I guess that goes without saying—we've been walking together for twenty-seven years. And on a trip like that there's bound to be variable types of times.

The author in the San Rafael Swell

For instance, before Jennifer was two years old I stabbed her in two places. Changing her diaper one evening, I cleaned the pink crevices of her flesh, powdered her bottom, and then slid the safety pin neatly through a fold of skin at her hip as I clipped the corners of the diaper together. Only her screams saved me from leaving her stapled to the cotton. I remember, as I tried to unstab her, that the blood on her leg and the two neat punctures in her skin made me tremble with anger and something else, something I'm still unsure of. Since then, I've paid other people, physicians mostly, to stab Jennifer.

I settle myself on a timber spanning one of the kivas, content to watch in the shade of the overhanging sandstone while Jennifer works with a meticulousness that would shame most archaeologists. She's slight, with dark brown hair clipped closed to her head and deep brown eyes, just now sparking with excitement; cutoff jeans, a T-shirt, an old pair of tennis shoes, and a tattoo on her left ankle. But mostly, she's a twenty-seven-year-old woman too rarely allowed by me and others to be a child, even while she was.

Jennifer has accepted a position in the graduate program in philosophy at the University of Minnesota. She is to become a genuine doctor of philosophy, and this is our last trip together for a while, maybe a long, long while.

It was hot as we worked our way down the juniper-choked chute that led into this canyon, hot and then surprisingly wet for July. But here, beneath a cantilever of petrified sand, the air is cool and the rock and dust dry as lizards' breath.

"Look, here's a bead!" she cries. "What do you suppose this material is? Could be bone, or maybe a carved acorn shell [there's Gambel oak growing high up on the opposite side of the canyon], but I think it's bone. What do you think?"

I think that at times Jennifer reminds me so much of her mother it makes my chest hurt. I think the only thing worse than Jennifer's continuous questions would be none of Jennifer's questions. I think reality is an illusion. I think Charles Darwin was only half right. I think there is nothing more fearsome than dawn over the Utah desert. I think my daughter is beautiful. I think no father ever found the right place to put his love for his first female child. I think Jennifer deserved a better deal than the one the cosmos carved out for her. And I think this universe is the oddest, most nearly perfect, and possibly cruelest damn place imaginable.

That's what I think, but what I say is: "I think they must've had a lot of time on their hands, or been content with damn few beads around their necks. Whatever this is, it would take some time to make a string of them."

"Well, they probably did have a lot of time. The men were usually away hunting or, from the looks of things here, smashing pots and scarfing corn, and the women were left home with the crops and the kids. Time to burn," Jennifer says.

I left home when Jennifer was six. Her mother and I never found a common place—Nancy was kind of vertical rock and old wood, I was sort of scorpions and warm water, and we were twenty years old when we married. We divorced in 1972, six years later. But the razor of separation didn't cut cleanly until a year after that, when I finished graduate school and headed off to southern California. As I said, Jennifer was six, Patrick was four. For the next six years we spent summers together, me and the kids. But of course, things were never again the same.

"You think," Jennifer asks, "I could take a couple of these things with me? There are so many of them here."

"You'll have to make that decision for yourself," I offer in the tradition of the harsh and unequivocating father I established years ago. But Jennifer knows all the issues involved.

The Anasazi pulled up stakes and left here sometime before A.D. 1300. No one knows for sure why. Some say it was drought that drove the Ancient Ones into northern New Mexico and Arizona—the way the dust drove Oklahoma sharecroppers to California. Plausible, I guess. But, it might've been renegade space aliens that gathered these people up and carted them off to the ice caves of Jupiter. Might have been—because like I said, no one knows for sure. I like that.

Whatever the reason, the Anasazi sure as hell didn't feel obliged to clean their apartments, garages, lofts, barbecues, or kivas before they pulled out of town. This place is a mess—potsherds, corn cobs, beads, bones, basketry, burned logs, bricks, mortar, and arrowheads are everywhere.

I leave Jennifer with her dilemma and move to the far end of the site. A nearly intact kiva still stands here—if stands is what kivas do. All of the logs that formed the roof are in place, and only a few holes have opened between them, just enough to light the circular pit below. The benches where the men sat, the fire pit, the flue are all still in place and nearly like the day the last log was laid and the first fire built. I move to the front where a piece of wall has given way and stare into the gloom.

The smell of piñon and juniper smoke fills my nostrils, and a picture forms inside me, a picture of this place in an old December—the riverbed below and the far slope snow-covered, deep green, immaculate white. Snow is still falling and smoke is curling from each of the kivas. The men inside are heavy with their own thoughts and the resinous smoke that swirls around them. Sweat weeps from their bodies. I wonder at those men's thoughts—thoughts of corn, ravens, pale ghosts among the rocks and trees; the knife edge of fatherhood, the dark pool that is woman.

Jennifer, steps around to where I am.

"Isn't this place great!" she says. "How do you suppose they made the walls so smooth?"

I am so deeply pleased to be here with her, here in this ruin, this ancient holy place, so much so that Jennifer's excitement keeps me on the verge of tears.

"It is incredible isn't it?" I offer. "The kivas, the bits and pieces of lives now seven hundred years dust. I wonder what it was like in its prime."

"They sure had lots of corn, corn and pots. Look here's another piece of bone," she says as she lifts the skull of some sort of rodent from the thick dust at the base of the wall. "Do you suppose they ate these?"

"No, I suspect they herded them, though—used their pelts for clothing and their bones for spear points and jewelry, traded necklaces made from their claws for tobacco and horses, made small rounds of cheese from their milk."

"Wow, how fascinating. Maybe we should get you out of the sun."

"Alright," I say, "I'll stop" (I won't). "But we probably should start back."

"I hate to go," she says. "But then, I'll hate to go no matter how long we stay. It would be nice to be back before dark. Alright, one last picture. How about you in front of the kiva. Yeah, like that. Good. Okay, I'm ready."

Our eyes make one last trip across the dusty face of this magical place. A last plume of Anasazi dreams blossoms behind our eyes, then, like caterpillars, the dreams crack into smoky butterflies and flutter off across the rocks. I shoulder my pack, Jennifer picks up a stick for company, and we start back.

One day, when Jennifer was eight, maybe nine, I left her alone at my apartment with an ailing Shih Tzu that I'd inherited from a girlfriend who threatened to have the dog killed unless I took him. There was nothing wrong with the dog, the girlfriend was just bored with responsibility. I liked the dog alright. But at that point in my life I was too busy with science to be much bothered with a dog. The dog knew that. Jennifer called me at work that day to tell me that Snuffy was becoming less and less responsive to her ministrations, pets, and cuddling. Busy with mouse-cell-surface antigens, I asked Jennifer to take the dog to a vet, who was about two blocks from our home. Jennifer did, gladly. The vet said he wasn't sure what was wrong and wanted to keep the dog overnight. Next morning, Jennifer called me at work again—this time to tell me that Snuffy had died. Renal failure. Yeah, sure—terminal disappointment in humans wasn't an acceptable diagnosis then either. I called the vet, told him we wanted the body. He seemed surprised.

I picked up Snuffy that afternoon, wrapped in a clear plastic bag and hard as water ice from a morning in the vet's freezer. The dog didn't look disappointed or angry or accusatory, as I thought he would. He just looked a little wretched from the poor haircut I'd given him a few days ago, and done—you know, done. Like he'd had some task to finish, and now it was done.

We buried him under a pine tree I'd planted behind my apartment (a Christmas tree I'd shared with the woman who left Snuffy with me). I dug the hole while Jennifer and Patrick watched. They cried, then I cried. And then, without a word, Patrick and Jennifer went into the house and returned—Jennifer with the last of Snuffy's dog food, Patrick with the dog's pillow and stuffed toy. They laid Snuffy's things carefully around him inside of his grave. Then I shoveled sandy southern California soil over Snuffy, Jennifer made up a little prayer, and we walked back inside.

Jennifer never said a sour word to me about Snuffy's death. But I can't believe it raised her opinion of me any.

The last mile or two of the trip out of this canyon is steep, dead steep—50 percent, maybe 60 percent grade—some of it across slickrock bowls gritty from a hundred years of pulverizing. Jennifer doesn't like this stuff—I know she doesn't. She didn't like it on the way in, and she doesn't like it now. But she doesn't complain, just inhales sharply at times, mutters some to herself, and grunts it out. Don't get me wrong; as I said earlier, we've had mixed times. When Jennifer was completing the trial of puberty, none of us could get along, no matter how we tried. When Jennifer finished high school, I asked her to move out of the house. It was a shitty time. But Jennifer packed up and moved off to Maine for about two years. Ever since we've been friends, mostly good friends.

At the top of the chute we've climbed, Jennifer stops to look back at what we've covered.

"I'm actually rather pleased," she says. "I didn't think I could do that, that hill—or cliff, more like it—but I did. Not bad. We did it! Do you think the Anasazi came in and out the same way?"

"I'd guess they did," I say, thinking of where we'd come from, where we were headed, thinking of men and women and children seven hundred years ago, seven hundred years from now, and pretty pleased and impressed myself. I didn't think I could do it either.

On the way back to town we stop at a drive-in restaurant for dinner. Jennifer orders the frybread, I order a bowl o' red. The restaurant is nearly full, an interesting blend of southern Utah ranchers, Navajos, and sunburned tourists like ourselves.

"I'm so pleased we found those ruins today," Jennifer says. "It was just grand."

"Me too. It looked for a while like we might run out of time somewhere up-canyon from there. But, the timing worked out perfectly. What's frybread?" I asked, "I've never had it."

"I used to get it all the time when I visited Nancy in Chinle," Jennifer explains. "It's this wonderful fried dough that's filled with all sorts of things. It'll be interesting to see what it's like here."

A stray dog, grimy from days and nights in places without floors, has wandered into the gravel and cactus at the front of the restaurant. Jennifer is watching him. Two children waiting for their food have also spotted him. They slip away from their mother and out the back door. Once outside, they grab up handfuls of gravel and begin hurling them at the dog.

Jennifer bolts up from her seat, points and scowls at the children, raps hard on the window. The children suddenly switch to throwing rocks at the cactuses. Their mother notices none of it, finally collects their food, heads out the opposite door, and waits for the children to materialize at their truck. She yells their names, they come, clamber into the truck, and leave, trying not to look at Jennifer.

Jennifer is still seething.

"Why would anyone want to hurt that dog? And how could anyone ignore their children so thoroughly and allow them to do that? It's things like that that make me just lose faith in people, just imagine we are all doomed to hell and maybe it's just as well."

 I consider her remarks for moment. I have very similar sentiments at times but for different reasons. Mine have to do with the desert and the fecundity of the human species. But at times, I'm sure as hell just as short of reasons to go on.

"Maybe there's more to this than we see," I offer.

"Like what?" Jennifer asks.

"Well, I don't know, maybe there's a sense to this we just can't see with our eyes."

"I'd like to think not. I mean what sort of sense could hucking rocks at a hungry stray dog make? Jesus, that pisses me off!" Jennifer says, and settles back into her anger.

I'm stuck at this point. I want to offer Jennifer something to dull the edge of what has just happened. But this is the point where I always get stuck. What sort of sense does throwing rocks at skinny stray dogs make? What sort of sense does Bosnia make, or Union Carbide and Bhopal, or breast cancer, or Orrin Hatch, or leprosy make? What sort of sense does a dam in Glen Canyon make?

That night on the curb in front of our motel in Blanding, Jennifer is smoking a cigarette, I'm working on a cigar, and the locals seem to be enjoying the show. I've been expounding my personal philosophies to Jennifer, something about cigars makes me do that. Finally, I come to the only place I can.

"Appearances to the contrary, maybe this is a perfect world in which all things are happening perfectly."

In spite of her degree in philosophy, Jennifer tolerates my ramblings rather well.

"You should read Leibnitz," she says as she exhales tendrils of cigarette smoke. "You have more in common than you might think. He just got a bad rap because of all that Nazi stuff."

"I will," I say, thinking of the ease with which Nazi stuff arises in men and wondering if I'm telling her the truth.

Having said that, that "perfect world" thing, even having said it out loud, I don't feel any better about anything. And the haze that boiled up when the rocks hit the dog still hangs over our evening.

For the first time since Maine, Jennifer is going to be living in another state, a state that seems at the moment to be just this side of the Crab Nebula. And this trip is nearly done.

She looks so small, so fragile—her knees nearly to her chin, her hands vesseled and curled like a woman's hand. She looks twelve again—well of course she doesn't, but she does. She looks like she might crumple like chiffon if squeezed too hard by the clutching parts of this planet, if forced to watch one more dog stoned by thoughtless children. Of course she wouldn't, but she looks like that to me, this moment, and it makes saltwater well up below my eyes.

When we lived in California, Jennifer would rise before all the rest of us each morning and go down to the beach to collect shells. She loved to see the things the sea had tossed up during the night. Some she kept, some she left behind—"For the people that can't get up so early," she told me once. "I always leave some of the good stuff for the ones who come later."

"Seems to me if they don't get up early enough, they don't deserve to find much left," I'd say to her.

"I know, that's why I think it's neat to leave things for them."

I liked that. But I wonder, late as I've come to this place, whether I'll find any limpets or mussels, turbans or razor clams, scallops, periwinkles, or good pointing-and-digging sticks once Jennifer's gone for Minneapolis.

I've spread a map in the gutter in front of us. We're trying, among other things, to decide where we might go tomorrow, our last day. I'm trying to choose between Grand Gulch and Dark Canyon. Jennifer is mostly bored with this consideration, until her eye lands on a spot just to the west and a little south of us.

"Let's go there," she says, and pushes her finger onto the map.

I look at it. Damn few contour lines, no obvious features.

"Why there?" I ask.

"Look at the name," Jennifer says, brightening some. "How could you have missed it?" Then I notice scrawled across the low hills and shallow canyons the name "Cheese and Raisins."

"A place with a name like that just has to be interesting," she says. "That's where we should go tomorrow."

Tomorrow, then, Cheese and Raisins it is. Tomorrow, that's where Jennifer and I will be.

Dumpster Diving

What was it I should have done,
that I didn't? What prayer or curse
should I have spoken aloud, but mumbled,
out of embarrassment?
        Or maybe it was the woman, bent beneath

a cart of rags,   or the man with the tobacco-stained face,
that I should have loved
but couldn't. Maybe that's how I've come
to this place with nothing more than a gnawing child

twisting inside of me. Where is the voice of god now, where
are the grisly archangels and the slouching beasts?
Where is the holy water and wafers of colored light
filled with the sufferings of St. Stephen?

Not even a film of ash or grit
where the flame once blazed, not even a spot of oil
where the rows of glassed-in candles guttered and failed. Just
the void, just a mind-blistering fear,   just the smell of burning trash.

A man in a woolen cap, culling, with his spatulate fingers,
aluminum cans from the green garbage bin, stops
to wipe a bit of rancid pickle from his wrist and notices
        me there. He smiles a flop-eared smile,

then reaches further into the watermelon rinds, the egg shells
butter tubs, and bacon grease, the overripe peaches and pears, the rotting
pancakes and pork loin, and pulls a doll's head from the sludge
in the bottom of the bin. He clears the slime from its eyes

and its lips, then glances at me. "For my daughter," he says,
        as though I'd asked.
Then he wipes his hands across his wrinkled pants, drops
the doll's head into his pocket, gathers his cans, and walks off.

For as this appalling ocean surrounds the verdant land, so in the soul of man there lies one insular Tahiti, full of peace and joy, but encompassed by all the horrors of the half known life.

—Herman Melville

Ravens

This morning two flattened mule deer, one truck-twisted cow elk, one dismantled skunk, and two putrid porcupines greased the shoulders of I–70 between the Eisenhower Tunnel and Vail Pass. That's unusual, even for I–70. Curious. But it wasn't the number of dead that unsettled me this morning. Nor was it exactly the absence of life that was eating at me. After all, there was that blue heron, stick-still in the marsh near Milikan, those three magpies working an unrecognizable chunk of roadkill near Gypsum, and two golden eagles fishing near Green River—not to mention the various bits of wildlife (or wild dead) in the yellow and brown clumps baking on my windshield. Nope, what was chewing me up this morning was the fact that I hadn't seen a single raven this day, not even a crow (as if I could tell the difference), not even way off, not even one. Why should that bother me? Let me tell you.

It seems inarguable to me that the past lays the trap for the future—an ordered universe driven by fixed forces must change in orderly ways. Call it karma, if you want, call it Callahan's first law of universal inevitability. It doesn't matter. But the implication is, if we can predict the probability of finding an electron a given distance from a proton (and we can), we should be able to predict the likely outcome of, say, a Knicks-Lakers game. It's simple mathematics. Thousands, maybe hundreds of thousands, of people share this vision with me every day.

Albeit, the vision fails most of them, as you can tell any day by the number of people trying to hitch a ride out of Las Vegas. But I think these failures occur not because of any flaw in my logic (as if I needed to say that) or the organization of the universe. I think these folks fail because they don't pay attention to enough variables. Generally, I'm kind of anxious about the future, so I've been trying to pay attention to more variables. It's important to start small here, or the possible combinations can overwhelm you. So at the outset I'm evaluating the "omen potential" of only those things that change while I'm actually observing them. Generally, this rules out geologic events such as continental drift, orogenies, and the Republican Party—undoubtedly important and future-focusing events, but just too damn slow to be of much help in deciding whether to have the cheeseburger or the fish sandwich. And in fact, because my attention span isn't what it might be, this time-factor thing pretty much limits me to observations of so-called living things for predicting the future.

Initially, I've settled on animals because, well, because they move fairly fast, and I don't have to watch for very long before something changes. And guess what? It works! Well, sort of. For Macbeth, it was his witches; for me, it's ravens. That is, generally, I've observed a reasonable correlation between the number of ravens I see and the probability that the immediate future will not involve my death or anything else really nasty. As a consequence, I pay more attention to ravens than I used to.

So I was uneasy about the aforementioned lack of ravens as I neared Utah Highway 24 and the turnoff to Hanksville this morning. So uneasy, I guess, that I missed the road completely and drove halfway across the San Rafael Swell before that bit of internal sense, the one with the maps and compasses and astrolabes, began banging on the backside of my forehead.

I pulled over and looked at the map on my car seat. The Best Western version of Utah showed no discernible feature, road, or milepost between Green River and Salina, about a hundred miles. I don't know who these maps are intended for, space aliens maybe, or SST passengers—there might be enough information here for people sipping Jack Daniels and headed for Tokyo aboard Air France at about 45,000 feet. But for humans and other surface creatures, these maps aren't much use. They're even too slick for toilet paper. So anyway, there was no way I could fix my position relative to anything other than California maybe, and certainly not Utah State Highway 24.

Wherever I was, though, the view to the north was worth the trip. Across what I figured to be the Jackass Benches, lay the rust and blue wedge of the San Rafael River Canyon. And all around were the deep rills and gorges of the Swell itself, a stone bubble belched from the iron-rich soup of the Utah desert 70 to 80 million years ago during the Laramide orogeny, and then splintered like old bones by shifting crustal plates and the sheer doggedness of time and water. Overhead hung that cloudless cobalt Utah sky.

Since everyone pretty much has to be somewhere, I figured this was an okay place to be, compared to, say, Tulsa or Aspen. But the fact remained, this wasn't Lake Powell. I returned to my truck, and as all six cylinders fired the two turkey vultures—working the heat-driven vortices above—moved off disappointedly. In deference to the gnome inside my head, I pulled back onto I-70, looking for a place to turn around. A mile or so later I found it, obtained the appropriate authorizations and made a U-turn.

Seventeen miles back the way I'd come, the flagwoman, Uintah perhaps, told me where to find Utah State Highway 24. Out of kindness I suppose, she explained that people missed that turnoff all the time. I appreciated her words but suspected they were spoken more for effect than honesty, especially after I saw all the signs and bridges and everything around that turnoff. As I waited for the skiploader to clear the eastbound lanes, I couldn't help but notice again, no ravens.

I have the boat along on this trip, a last float, a farewell to one more possession. I think it was the neopostmodern, deconstructionist philosopher and guitarist Leo Kottke who said, "I used to hate cars, so I bought a boat. . . . Now I hate boats." Leo knows his shit about watercraft.

The nature of boats is such that any trip that includes a boat is likely to have more memorable moments than a trip without a boat. This trip was to be no exception.

An hour later than expected, but in fine fettle, I arrive at Bullfrog Basin. I back the boat and trailer down the ramp from the very top. No shortcuts for this mariner—a fact you are unlikely to appreciate unless you've seen the ramp at Bullfrog when the lake is sixty or so feet below high-water mark. I dump the boat off the trailer into the water, then start the outboard by jumping it from the auxiliary battery—the primary's dead (a bilge pump switch I never find time to fix). I move the truck to one side, then hop aboard and motor toward the gasoline dock and my rented slip. It is good to be on the water, to be, again, with the river. The air is warm, the lake coral blue and cool, and for a change my boat is running smoothly. I stifle any thoughts about what lies below this water.

Abruptly, as I round an intervening spit of red stone, a twenty-foot fiberglass boat bursts into flames—still tied to the dock—and two people scramble off the bow. Oily ropes of smoke coil off the stern, then the gas tank lets go, and orange flames fill the boat's belly. The windshield puddles.

Glen Canyon Dam, Page, Arizona

For long moments the heat is too much, and everyone just watches the boat burn, watches slack-jawed, thinking of their own boats, their own gas tanks. Finally, two men in a Park Service skiff get close enough to get a hook and line onto the boat and pull it, still burning blackly, away from the dock. The rangers pump water from what looks to be a two- or three-inch hose, but it makes no difference. The fire just keeps eating at the glass boat until most everything above the water is gone, until nothing but blackened and boiled bits of hull and motor and a piece of life jacket remain. The air is rancid with burned plastic and gasoline. The rangers tie what remains of the boat to the opposite shore and motor off.

I work my way past the wreckage and pull up to the fuel dock. A man in a Park Service uniform offers to fill my tank. While he's pumping a mixture of oil and unleaded Chevron into my tank he tells me a woman—partly, maybe wholly drunk—lit a cigarette and, laughing, tossed the match over her shoulder. Casually, she looked back to see where the match had landed. She shouldn't have. She saw the match still burning near the gas tank. Even drunk, her body responded. Reflexively she reached for the match. She reached, and the gasoline reached back, reached back and took her face off, then ate her boat.

Surprisingly soon a helicopter settles near the dock, and the woman and her companion are whisked off to Price or Grand Junction. Perhaps the people there can help her with her face. But she won't finish this day with the face she began with. There's damn little for certain in this world.

I tie up my boat at the slip and climb out onto the dock. Overhead, from nowhere, three pairs of ravens appear—six black-beaked oily-feathered flesh eaters—working the warm air. Shaken, I walk the two miles from my rented slip back to the launch ramp for my truck. It helps to have my feet on the bare rock, solid, certain, gritty. But I refuse to look at the black birds overhead, refuse to even consider the questions they're screaming at me. The birds cry angrily and pass off to the south.

I sit in my room for an hour or two trying to assimilate what I've seen. I can't. So I gather my things and walk back to the boat. Still mildly nauseated, I push the boat at full speed across Bullfrog Bay. The wind and the water spray over the bow help. At the main river canyon I turn downstream and motor up the first side canyon I find—a short one, deep, abandoned. I shut off the motor, strip, and slip over the side into the river. The water reaches for me, slides its broad cool arms around me, and holds me. The nausea and the stench of burned boat pass. The river whispers to me—words I knew once but have forgotten—over and over. My mind empties. For now, it is enough to be just one living man up to his nose in this vast river. The water gathers me.

As it turns out, it is much easier, especially when naked, to get off a boat than it is to get onto a boat, from the water anyway. It's about the same from the dock. It was good I hadn't thought to raise the outboard. Otherwise, I might be writing this in falsetto. I sit drying in the warm evening air, surrounded by sandstone cliffs once a thousand feet or more above the river. A swift spirals down from above and disappears into a hidden nest. Ravens call from somewhere deep in the rock, and their cries ricochet hollowly along the dark face of these old walls. And water, come from God knows where, drips from a mossy slot overhead. I listen for a while to the ravens, the water, listen to the thump of my heart, and the blood thrumming in my ears. I shout out loud to hear my words spoken by these stones. And then it is quiet again, quiet like it was before there were words, before there were birdcalls or moonlight—quiet, like when there was only the water slapping at the stone edges of the silent dark world.

The next morning, after a fine long walk along the solid shore of this bay, I start gathering gas cans—I had underestimated this water. The nearest marina, Dangling Rope, is over sixty miles from my boat slip. Even with simple math that's 120 miles, give or take, round-trip. Let's see, my boat carries ten gallons of gas, burns five per hour, and reaches a top speed of twenty-three miles per hour. Likely I'll require more fuel. (Thank God for my schooling; no telling what might happen to some poor uneducated sap out here.) I buy some maps, some baloney, bread, beer, a lot of gasoline, some more beer. Evening already colors the sky. Time only for a short boat trip around Bullfrog Bay, then early to bed.

Finally, on the second morning, an opportunity for some touring. I motor the canyon from Bullfrog to Dangling Rope—with a side trip to Rainbow Bridge, of course. The Anasazi who first lived and died here, as well as the Navajo who live and die here now, think this rock is sacred. You only have to see it once to know why. The firmly muscled forearm of the earth itself is risen here in colors of flesh and sand, risen from its stone shoulder to bolster the sky. And it must wear on the sons and daughters of those first people to watch the bloated river disgorge sunburned men and women who shout at their children and at one another, who leave their trash and their urine here in the shadow of the great stone span. It must gouge at them something fierce.

Evening near Bullfrog, I stop for a little fishing and catch a little fish. Then an early dinner and bed.

On the third day I rise with a dead serious need to hike, to get off the water for a while and use my hypertrophied thighs and calves. I head off across the bay again, down the main river canyon, through a fist in the rock, up a minor drainage, and find a place to beach the boat.

Barefoot, I step from the bow to anchor my boat and deftly fill my feet with about 400,000 tiny barbed spikes from some sort of wretched little plant that has colonized the beach here. Hopping and cursing, I manage to get the boat anchored first, and climb back in. I begin removing the little needles from the soles of my feet. Goddamn plants anyway. This nature thing can be taken too far some times.

The Colorado River where it enters Lake Powell, near Hite, Utah

Most of the barbs are as small as the finest of cactus spines. So after a couple of minutes I give up. I pull on two pairs of socks for padding and lace my boots tight. I step from the bow of the boat to the earth—it's bearable, but painful enough that I probably won't do that again.

The small slot I'd beached in comes to an abrupt halt about thirty feet from the boat. But the headwall looks scalable. Part way up, in a crevice off to my right, I find four deep green datura. Two of them are blooming in the cool dark before this morning's sun. Blossoms big as my fist, eggshell white with yellow veins and horns, anthers thick with pollen, anxious for the thighs of bees.

The guidebook says of datura: "All parts are poisonous, have killed cattle, sheep, and children." Which is true. But hell, who among us hasn't at least wished to kill cattle, sheep, and children? The guidebook doesn't tell the rest of it, though, doesn't even mention how datura has long been used for things other than killing ruminants and miniature humans.

The few cumulus that have been waltzing with the sun dissipate in the day's rising heat. And abruptly, I'm once again standing in direct sunlight. Time to move. I climb out of the little canyon and set a rock cairn where the canyon meets the stone flats above. Then I head off toward where I imagine the river to be. Nothing but Navajo sandstone, burnt and blistered, spalling white and brown and black. Wave after wave, petrified by their own weight a very long time ago. Watching the whiptails watching me, I move out across the banded stone. I love walking on slickrock, love the flat slap of it against my boots. I don't know why, but it beats any pavement, any lawn, any decking, any bowling alley I've ever set foot upon.

Near the top of the mesa, there are hundreds of gray and brown concretions that look like little "dinosaur eggs"—an inch or two across, left by erosion standing on finger-length stalks of ancient mud—teed up, as it were, for the long irons of Thor and Loki. They're everywhere, like petrified roe or marbles or small stone testicles. Some split, by ice I guess, are showing light brown yolks rimmed with a darker grit and a rough brown-to-black shell. I pocket a few of these eggs—for my wife, you understand. I wouldn't gather rocks for myself in such places.

At the western edge of the mesa the land just falls away. The river spreads blue-black in both directions, speckled with sunlight, wrinkled by the wind. Five hundred feet, more or less, slightly overhung, to the water. From my stomach (in these places the wind is always at your back and the ground always slopes toward, never away from, the precipice) I look down the fractured and varnished wall, past the hideous water stain at the bottom, into the river. I can't help but think of the view before water filled this canyon, think of the shape of the river bottom, another 200 feet below, its sandbars and gardens, rapids and falls—can't help but think of Powell in his wooden boat with his only arm. I think about the silt, the dam, the long once-free run of the river. I wish out loud that things were different than they are. But that's just selfish, self-indulgent. Lots of people wish things were like they used to be, even Republicans. Nobody gets it. Except maybe the water.

It's time to eat. An apple, a plum, some heat-limp Swiss cheese, and the last of my water. I envy the white-throated swifts and their knowledge of warm air. I wet my finger and press it to the stone where I sit. Then I stick the finger in my mouth. Grittier than I'd imagined, a little less spicy, tastes sort of . . . stonelike. For a while I watch the power boats frothing the water below. Then I watch the clouds frothing the sky above. Assured everything is as it must be, I take a long breath of the water-laden air, stretch my bones out, and nap.

On my way back to the boat I cross an old watercourse. A polished rut through the mottled stone with three large potholes scooped by water and wind from the flesh of the rock. The holes are ten to fifteen feet apart and about six feet deep, swirled with chocolate and vanilla stripes. Dry now. At the bottom of each tank there are a dozen or more "eggs"—testament to the fierceness of the rain here. A stone rookery, full of ancient ova carried here by rainwater to ripen in warm rock hollows. The last of the thunder lizards perhaps, doomed to an eternity of stone because their parents lost interest and left with the birds. I can imagine worse ways to spend the millennia—as Bob Dole, for example.

I find the cairn that marks my canyon and scatter the stones (each of which seems pleased), then work my way back down the sandy headwall of the little canyon. Once more I pass the datura, its blooms shriveled now by the photons boiling off a thermonuclear furnace some 93 million miles distant, a furnace casually changing hydrogen into helium, and helium into the dark matter of life.

Back at the boat I take off my clothes and sit heavily in the waist-deep water wearing only my hat. The river here is no more than a lime-green swirl over red mud, a liquid question over stone that is once again sand. I watch where the water laps at my overhung waist, watch it rise and fall slowly across marks left there by another smaller hotter sea.

As I sit in the tepid water I think of the burned woman, her new face forming under red gauze. I think of the datura, white as the Holy Spirit, green as fecundity—feral weed full of pretty pictures. And I think of those stone eggs boiling up there in their stone cauldrons. Sunlight and rain clouds twist slowly overhead.

Maybe it is just too complex. Maybe there are too damned many variables—the angle of the sun, the wind on the water, the swirl of the clouds across the moon, the snake's shed skin, the dragonfly's wings, human kindness, the broken beer bottle, the single blade of grass bending in the wind beyond the fingers of my left hand. Maybe it's hopeless.

Two ravens rise in the thermals above, carving black helices in the blue air. Wind spreads down-canyon like warm linen, ruffles the water, then lopes out across the lake. Sometimes, though, I'm pretty sure that even all of it wouldn't be enough. Even if I could name every piece of this universe and account for its whereabouts at the instant of my decision, I'd still be wrong about 75 percent of the time. I'd lay money on it.

Raven's Dream

The day before yesterday, just after noon,
people
began falling from the sky.

A hole opened
between two copper-bottomed clouds,
and people

began falling out, falling
straight down
onto a section of plowed earth.

There were men in overalls
and sweat-stained ball-caps, women
in flowered aprons, and

children in denim with dogs and cats.
And where they fell, the blood of each ran
from their shattered bodies into the black,

newly broken clods.
Then came rain, picking at the earth.
Softly, at first, but soon, fists

of rain pounded the field, and water
burst in foamy rivers down the rows, pushed
into the galvanized culverts

beneath the roads. And suddenly,
where each person's pulp had reddened the soil,
from where each ruined body had lain,

there rose a shimmering stalk with tasseled ears,
and kerneled heads beneath broad green spears.
Then the rain stopped. The iron clouds crumbled,
and beneath the new-born
sun,

raven
came, ruffled black,
and fed upon the ripe yellow corn.

It is not down on any map.
True places never are.
—Herman Melville

My Canyon

Snowflakes big as spiders, crowd the light from the sky, and the wide red mouth of the canyon has turned sugarplum white. The river below inhales and exhales, and the snow is sucked downward or blown upward, then, for no apparent reason at all, torn sideways. Beyond the snow a fog is growing, thick like old cream, swelling, dwindling, revealing then concealing visions of what once was, and what yet may be—but certainly not in my lifetime.

Pat and Steve Larkin near a cloud-filled Grand Canyon, Arizona

My sister, Pat, her husband, Steve, and I huddle beneath a thin roof of Entrada sandstone and peer into the depths of the last glorious stretch of Glen Canyon—the precious few miles left below the concrete wedge of Glen Canyon Dam. The emerald ribbon of river far below the fog gobbles up the snow spiders quick as they fall.

Just now we're eating turkey sandwiches and alternately arguing over who was supposed to have brought the beer and bitching about the dam just out of sight to the north. It's a beautiful spot, a glorious day, a good argument.

I move closer to the edge to watch the snow spiral into the depths below. The flakes beckon, and I am drawn to follow them. Each crystal, born perfect only a few feet above me, is extinguished as perfectly a few feet below—an instant of frozen loneliness between the families of sky and sea.

I stifle the urge to follow the flakes, step back to the overhanging stone, and settle again with my sandwich. From here you can see, or you could before the snow swallowed the light, one mile or so of the only fifteen left of Glen Canyon—vertiginous, underslung halls of red stone cradling a once violent green serpent. It must have been splendid, this canyon, and all of its tributaries, so full of fertile clefts and grottos that Powell himself named it Glen. Splendid—twisting across near two hundred miles of what is now Utah—an image of the river itself, cast in stone as the brittle crust of this plateau rose up like warm bread. Splendid, but mostly covered now by the wet drape of the river. And this Saturday, somehow, what remains must be enough.

"I heard Echo Park was just a diversion, a head-fake. They intended all along to build the dam here," says Steve, reading my thoughts.

He's referring to the story that the 1950s Bureau of Reclamation and its fiery director, Floyd Dominy, supposedly first targeted a spot near the confluence of the Green and Yampa Rivers in Dinosaur National Monument as the site for a dam that would ship power to the highest bidder from Minnesota to California and water to the thirsty from Salt Lake City to Los Angeles. But people, lots of them, had seen Echo Park, a dramatic spot inside a national monument that would be flooded once the dam was complete. So the public outcry, especially as amplified and orchestrated by David Brower and the Sierra Club, was so great that even BuRec decided to look elsewhere. I don't know if the story is true or not. It's certainly true that David Brower pretty much by himself blocked the construction of the dam that would have flooded Echo Park. But my recollection is that Echo Park and Glen Canyon were among the original four dams approved by Congress as part of The Colorado River Project in 1956.

Regardless, when Floyd and his cohorts proposed Glen Canyon Dam at the end of a little known stretch of red rock canyon "useless to anyone" near the Utah-Arizona border, the public politely looked the other way. If there truly was a choice, I'd rather we'd lost Echo Park. But in spite of my wishes, in 1963 Glen Canyon Dam began backing water 186 miles into Utah and flooded forever hundreds, perhaps thousands, of rarely seen places—places with magical names like Cathedral in the Desert, Mystery Canyon, Music Temple, and Labyrinth Canyon, places full of flowers and ferns and the bright red flesh of the earth.

"Maybe even five years later, certainly ten, it would never have happened. So it goes," I say. "At least Floyd didn't get the ones he wanted at Marble Gorge and Bridge Canyon on the Grand."

"Yet," adds my sister.

"Good point Patricia," I offer, "but dam cynical."

"Ugh."

"Dam good sandwich too, Steve."

"Well, thanks," said Steve, "I've somewhat of a feeling for fowl."

"Yes, you do. You certainly do," Pat responds cryptically.

Pat and Steve would spend all of their time on this river if they could. As it is, they spend the better part of three months between every June and September—when Pat is free from her teaching—on one part or another of the Colorado. They're boat people, and pretty good from what I hear. They're both somewhere around fifty. Pat has some of the same genes I've got, brunette hair cut short, streaked with gray now, piercing brown eyes, and a face creased by a million smiles. She's short, a little fierce, with the damndest thing you ever saw—prehensile toes. That's right, like a New World monkey's tail, or an elephant's snout. She can pick up dimes with those things. Incredible. Steve is tall and lean, equally quick to smile or laugh through his thick beard. His flint blue eyes behind wire-rimmed glasses always seem up to some mischief of their own. Good, honest people, stained by their years on the river, by their lives in the desert. They work when they must, boat, hike, and lie out naked under the blazing stars whenever they can.

The snow is still falling heavily, but now the wind has kicked up, and the snowflakes are moving mostly horizontal. I notice that Pat has settled into a near continuous shiver at this point. I pull my down vest from my pack and pass it over to her. I take a lot of crap for my big day pack and all the shit I just won't go hiking without, but sometimes it's a help. I demand some acknowledgment of this.

“Thanks,” she says, then adds, “But it’s on account of you I’m here, remember?”

“You could do worse,” I quickly point out.

“You mean worse than you, or worse than here?”

“Well, either, . . . obviously.”

“I’ll grant that it’s a fine view,” she says, and then our conversation stalls again for a bit. Pat’s teeth settle to a bearable, for me anyway, chatter and our focus moves back to the canyon. Two things I’ll miss when I’m gone—words with friends, and silences with friends.

Yesterday I was with my father, Patricia’s father too. He and I drove from Kanab to St. George to see a urologist there. That’s really why I’m here.

It was snowing heavily when we left Kanab—big wet flakes gathering quickly. Dad was driving, so I was left to watch the snow painting the pavement and gauzing up the sagebrush at the road’s edge. My mom was in the back seat with their “new” three-year-old cocker, Cocoa. I’d wanted to take my truck for its four-wheel-drive. But first thing that morning I’d locked the keys inside it—something I never do. So we were in my dad’s Ford, moving slowly through the muck toward St. George. My dad’s eighty-three, my mom seventy-nine. I don’t remember much of what we talked about, little things—the snow (“Thick as I’ve seen it”), road signs (“6 Miles to 3-Mile City”), the beauty of the mountains here (“if we could only see them”)—the usual stuff. My mother petted the dog.

My parents seemed much older than when I saw them in August. I rode next to my father and thought of all the places he’d driven me. The snow quickly daubed out one of those Jackson Pollackesque pictures it seems so fond of, and the miles slipped beyond the black Firestones beneath us.

We arrived in St. George with some time to spare. So Dad took care of some business and stopped by K-Mart to pick up some aspirin.

“They’re two dollars cheaper here than at home,” he said to me, as though an explanation was needed.

Then we drove to the urologist’s office. Mom used the restroom, then said she’d wait in the car. Dad walked her back to the car. Afterward, he and I went into the doctor’s office.

Between being scalded when I was three and all the usual childhood afflictions, there must have been twenty times anyway that Dad and I sat together in doctors’ offices, twice that number with my mother.

Each one of those offices was just like this one—a polished plywood rack of magazines (here supplemented liberally with Mormon literature), a large rectangular hole in the wall through which you could see a receptionist looking as though she'd just as soon not be bothered with whatever it is that's on your mind, and rows of patients' records in fat manila folders with pastel-colored labels. Thin gray carpeting hugs the floor, perforated ceiling tiles float overhead, and a heavy brown door leads off to tiny examination rooms. And throughout, the faint smell of isopropyl alcohol.

Same office, but this time it was *his* doctor, not mine, that we were waiting for. This time *I* had to know what the doctor was talking about, and this time, as my father always had, I had to act as though I knew what to do about it.

Until we walked into the clinic I hadn't even known my father was seeing a urologist. I'd thought he was seeing a GP, as he had the whole time I was growing up. I told him that.

"You told me once I should see a specialist when I have a special problem. I did that."

I never imagined my father thought enough of what I said to change the way he did anything. If I had, I would have thought more carefully about the things I told him to do. The physician and I were on equal footing, though: My father hadn't told him I would be there.

"I'm sorry about that, I just didn't get around to it."

"It's all right, Dad, we'll manage."

"The other times I've talked with him," he added, "I felt like I asked the wrong questions. I don't know why, I guess there's a lot I don't understand about this. I'm really glad you're here."

"I'm glad I'm here too."

But inside, I would have given anything to have someone else here, someone who had some idea what the hell he was going to ask.

There were two or three other people in the room. Dad and I spoke softly to each other. Dad was more comfortable with all of it than I was. His PSA was up considerably from six months ago, and there were three new spots in the radiograph of his ribs. But I wasn't here in this physician's office to help my father choose among potential cures. They don't even try to cure eighty-three-year-old men with prostate cancer. The urologist was just trying to slow the spread of the tumor until my father died of something else. So my job was to offer advice on how best to do that, how best to avoid the short straw today and keep my dad's death options open for a few more years. And I already knew that if my twenty years as a biologist were going to help at all, it would only be with the easy parts.

A nurse appeared at the polished brown door—crisp white pants, sweater (they're always cold), stout shoes, tired-looking eyes—and called my father's name. Dad explained to the nurse that I was there to discuss his treatment with his doctor. She looked at me and smiled a little uneasily, then without comment directed us to examination room three.

The author's father, Wayne Callahan, at The Neck, near Moab, Utah

In the end, my dad chose a hormone therapy. You can fool prostate cells like that sometimes, for a while anyway. But most importantly, Dad chose. He only wanted me to help him ask the right questions, as he said. Once he got the facts he needed he didn't even look my way, he just told the physician to give him the shot. I realized then what I was most afraid of. I was most frightened I would be asked to pick among the options and choose the course of my father's life. My biggest worry wasn't my father's disease, it was that I might fuck up. For what it's worth, I think my dad did the right thing, and I told him so. For what it's worth, I think I fucked up anyway.

Insights like that are useful, I suppose. Like the time one Thanksgiving I split my finger like a grape. For a moment, I could see the tendons and bones inside my finger—useful, but it hurt like hell.

It's still snowing, and it's still cold. The canyon is coated with an icy rime. Silence piles up with the snow while each of us sits silently. The canyon smiles whitely. Finally, Pat suggests we head back to their place in Big Water to warm up, get some food, have a drink. I could use all three, so I quickly second her motion and we start back to the car.

It'll be four months before my father's next bone scan. After that, we'll know if the prostate cells lodged in his ribs have down-regulated their receptors and closed up shop. I'll have until then to prepare to do better.

The next morning Pat's up early for school and Steve's up early out of plain meanness. So I roll out of my bag at a slightly more decent hour than usual. Pat heads off for her classes as Steve and I mop up the last of the burritos he pulled together for breakfast. We talk for a while after eating—nothing in particular, just talk. About nine o'clock, as the sky thickens into another gray soup, I decide to leave a day early to see if I can drive out from under this winter storm, maybe even get in a day or two of hiking north of here.

Steve, I know, would have preferred a day together. But you wouldn't know it, so quick is his adjustment to the changes. As quickly, he alters his own course for the day and settles into the new path. Rivers do that to people.

I remember only two things about the drive from Big Water to Moab. As I passed the Navajo Power Plant on Arizona Highway 98, a BB like a miniature brass SCUD missile slammed into my windshield and lodged there. I was so startled by the whack of metal against glass that I spilled about half the jar of orange juice I was sucking on. I slowed, looked around, but saw no one. So I drove on—with a BB planted firmly between me and the rest of the world to remind me, for as long as I have this windshield, of steam above coal-fired turbines, the stark beauty of Navajo Mountain through rifts of northern Arizona fog, and the odd bonds that form between people, the little reasons we have for loving or hating one another . . . planted to remind me of the small clusters of cells growing inside my father's ribs.

I remember that, and I remember passing beneath the long green tongue of Mr. Peabody's coal chute where it crosses U.S. Highway 160 and fills the bellies of railcars—railcars with no human hand at their controls—full of Black Mesa coal for the Salt River Project and the fires of the Navajo Power Plant.

Mr. Peabody's coal chute, near Kayenta, Arizona

And then I was standing beside my car near the mouth of a canyon I discovered after a busted White Rim trip with my brother Michael. On that day there was only time for a hurried reconnaissance, but it was enough to assure me it would be worth another trip, and that I would need a rope for that trip.

I push all the stuff I imagine I might need into my pack, take a last check on the sky, stash my car keys, and head off up-canyon. The route goes through a first canyon to a waterfall, then over a hogback into a second canyon, my canyon. Both canyons are heavy with water, so last May when I was here, mosquitoes—one of just a few of God's creatures (along with Republicans) that I have to try very hard to like—were competing with my biceps and quads for the quart or so of disposable blood I carry around. Not so today. This January day, as I work my way through the tamarisk and willows that choke the mouth of the canyon, mosquitoes are still just clumps of nucleic acid and glycoprotein buried in the mud beneath these iced pools.

The sun isn't shining, but neither is it snowing. And there are clear signs it's been warm and wet here already this year. Blades of grass are gathering green where I walk. Just the sight of them washes through me like a warm wave and cleanses some part of me that's been without a bath since fall. My breathing deepens, and I hear it bouncing softly back from the sheer stone walls of this first narrow little canyon. I negotiate a five- or six-foot dry fall, battle my way through a final bit of brush, and I'm at the end of the first canyon.

On the north side the canyon rises dead-vertical to form a hundred-foot-high falls, dry just now. At the foot of the falls is a plunge pool, maybe twenty feet across, surrounded by red sand. All of it is frozen now in the deep winter shadow of the falls. I kick a bit at the ice at the pool's edge and revel in the crystalline echoes as the ice shatters. I toss a rock or two onto the frozen sheet, where they lodge indifferently. Indifferent now, sure, but just wait 'til spring!

The south side of this canyon opens steeply to a stone ridge like a camel's back. I head that way, out of the shade near the pool. The air warms noticeably, and I stop to pull off my down vest. Where the stone first crests it forms a broad bowl—one hundred, maybe two hundred yards across. The view opens out here, and wave after wave of frozen stone rolls off to the east. The solid gray cover above me appears to be breaking up some. I stow the vest in my pack, grab a little water, and cautiously skirt the steep western lip of the stone bowl to a low ridge.

Across this ridge line the sandstone gives way briefly to a thin soil and sparsely placed pygmy junipers. I work my way down, toward a second canyon. The angle of this canyon is such that you can't see it at all until you're right on it—the hundred- (or so) foot-deep defile of my canyon. The air here is thick with juniper.

I work my way to and then along the rim of the canyon looking for a low spot, a place where the drop is short enough for a doubled-rope rappel. The stone feels incredibly good beneath my boot soles. Finally,

I find a spot that looks promising and work my way across and down some high-angle rock to the canyon rim. Perfect, no more than forty or fifty feet to the bottom, and a juniper tree has volunteered to belay me. I slip my pack off my shoulders, pull out my rope, runners, and harness. I pick three loops of red webbing that should be the right length to gird the tree for a rappel anchor. I loop the first runner around the rough bark of the tree, loop one end through the other, and then attach the other two strips of webbing with girth hitches.

Looks pretty good. I clip in a carabiner and lean back hard to check the anchor. Everything seems okay. Just a little harder, to be sure. Abruptly the whole thing lets go! I pitch violently backward, clenching the suddenly useless runners in my right hand.

One of the things I'm not very good at sometimes, especially times when I've other things on my mind, is judging distances, even short distances. As I'm hurtling backward I recall that the canyon edge was behind me, but I've no idea how far behind me. Instantly, though, I'm slapped hard by a rough fist of stone, knocking the air from my lungs and the hat from my head. For a moment I lie, genuinely taken by the depth and breadth of my stupidity. And then I get to my feet and turn to find my hat and reset the anchor.

Where I turn to look for my hat I find only air. Though I didn't make it over the edge, the placement I achieved on my reverse header was sufficient to launch my hat into the space above the canyon. My appreciation for my own foolishness ratchets up one notch.

I girth-hitch the runners back together, locate a spot farther from the edge, and again lean back hard. This time everything holds. I've no idea what I did wrong last time—girth hitches are not complicated knots. Oh well, it seems right this time.

I drop the rope over the edge, then run it through the anchor and pull it to its midpoint. Then I strap on my harness, put my pack back on, loop the rope through the descender, and lower myself over the edge. Everything, including me, does what it is supposed to do this time, and I work my way down the canyon wall.

At the bottom this canyon is relatively wide and filled with leafless willows and pale tamarisk. Elsewhere, large and small junipers rise deep green from the red soil. I unhook myself and stow my gear. I leave the rope in place, just in case. Both directions look inviting. After a drink of water I head up-canyon. The canyon narrows a little, and the walls spire up deliciously. Dead-vertical slabs of Navajo sandstone flat as Amarillo farmland and big as I can imagine. A little stream works its way down

Reflections in a stone tank inside my canyon

the middle of the canyon—water here, because of a more southern exposure, in its liquid form tirelessly scrubbing away at the canyon floor.

The canyon proves to be shorter than I had hoped. After less than a mile I can see the headwall and the boxed end, though I can't yet see the last of the canyon because of the undergrowth. Disappointed, I work my way through a final stand of willows and cattails. At the very last the brush stops abruptly, and I can see that the canyon forms a deep grotto overhung by wet stone. When I stop, the grotto fills with the sound of seeping water and a steady drip like rainfall in spring. The sun moves from behind the clouds. The air warms noticeably.

The overhanging stone walls are covered with a thick twist of ferns—some of them green even now, and what will later be monkey flowers and other smears of red and yellow and purple. Below it all in the middle of the cirque is a deep pool, thick with reeds at the near end. I work my way through the mud to the right side, drop my pack, and settle on a large flat rock. The air is a moist syrup of piñon and juniper, rotting ferns, cattails, old water, rich mud. Water seeps steadily from each of the fine cracks in the underlying Kayenta sandstone and flows slowly between the dry knees of the winter-killed reeds and grasses. And over it all there is a smell like linen damp from lovemaking.

I'm glad I was with my father while he considered his options. I'm glad he chose the hormone over removal of his prostate, or his testicles, or both. I guess I'm glad too for the glimpse inside myself. Perhaps I'll do better next time. Perhaps not. Fathers and sons are given an odd set of tasks.

The planet has turned far enough now to allow the sun's light to fall on the stone walls above and beyond the pool. On this windless afternoon, a perfect reflection of all that is above has formed in the pool. I stand and move to the water's edge to see it all more clearly. I have the very real sense that a sky-blue abyss has opened at my feet. And if I were to step from the rock now, it seems I would fall upward through this silent cirque into the blue vault overhead and then through the darkness beyond that blue, forever. The temptation to take that step nearly overwhelms me.

Instead, I step back. But my attention is still stapled to the pool's surface. Two ravens appear at the far edge, working their way along the canyon rim. In the stillness I can hear their wing beats, and the muted roar as the air splits where they dive and climb. Magnified and focused within this parabola of rock it sounds almost like breath, like the breath of the stones themselves. In a tangle, the ravens soar and circle, tumble and twist across the pool's edge, then slip beyond my sight. Silence descends behind them.

Only I and the empty reflection remain. And then I pray—a prayer I've prayed many times, a prayer to myself, perhaps. Grant me just one brief glimpse of the sense of it all, one insight into the mind of the machine that turns days into months and months into lives, turns boys into old men and rocks to dust. An instant of understanding, one wisp of why. You can take it from me afterward, if you wish. You can take all of it.

Silence.

I shoulder my pack and start back.

Vows

One naturally takes vows—
poverty, chastity, obedience.
Mostly it's poverty, though. And mostly
it's forgotten—the vow, that is.

Until one morning,
over eggs yellow as hen's eyes,
red beans and boiled coffee,
      we're speaking about a man
in a wheeled chair sowing corn
in the white-tiled furrows
of a nursing-home hallway.

And unexpectedly you recall the birds
in the forest where your mother is lying
unable, any longer, even to
swallow.   Then—as we finish our

biscuits and rhubarb jam—you speak
openly of your vow, your vow of poverty.
And the words you choose lead us
to the only place where anything is ended —

a field of corn at the end of a dirt road
with nothing left to us, save our broken limbs,
this crust of bread,
and the solace of our vows.

Because I could not stop for Death—
He kindly stopped for me—
The Carriage held but just Ourselves—
And Immortality.
—Emily Dickinson

The Dead Man Near Monument Valley

It was as though the great god of this universe had blinked, and in that instant when his lids were closed every movement, every sound ceased, except for the ticking of the asphalt as it heated in the morning sun and the slow drip of blood and gasoline onto the warm pavement.

Then all at once, everything—birds, people's voices, wind in the sage and juniper—began again. And there he was, facedown on the asphalt, dead, in a growing pool of his own making.

It must have happened only moments before. There were no police, no paramedics, and only three cars sat between me and the tractor-trailer rig sprawled like a broken mantis across U.S. Highway 163. I was somewhere between the awful spires of Monument Valley and the grim streets of Tuba City watching this man's blood work its way across the oily asphalt and puddle in the red dirt beside the road.

It was likely, though I couldn't tell from where I sat, that the man on the pavement was Indian—Navajo or Hopi, probably Navajo out here. His red Jeep lay mostly dismembered in the ditch forty yards or so from where I stopped. And he lay nose to the asphalt fifteen feet from the hood of my truck. Even from this distance, one look said he was dead, fresh dead maybe, but full dead. Someone had laid a towel over his head, or what remained of it.

The driver of the truck—fortyish, in blue jeans and T-shirt, with a three-day growth of stubble on his face—was talking to the driver in front of me, explaining in jerks and shudders how "The guy in the little red car was just out of control, pretty much . . . just out of control, completely out of control."

But all I could hear was a sound like a great steel locust, and all I could see was the back of the dead man—palms up, pink and smooth, a clean white T-shirt, and a hole in one of his cotton socks where his foot rubbed against his shoe. My mouth tasted of tin.

Monument Valley, Arizona

Four or five people were out of their cars, offering help. Somewhere, someone said an ambulance and the highway patrol had been called. There was nothing else to do. He—someone's father, brother, son, husband—was once-and-for-always dead. And all I could do was sit and try not to look while the sweetness of his life crawled thickly across the pavement into the mud. No thoughts, only colors.

Finally the pictures inside my own head snapped into focus and I realized I could pass by driving off the road, through the bunchgrass and beaver-tail cactus, around the wreckage of the truck, and back up onto the highway. Every place else seemed better than here, so I pushed my truck into low-range, and, with a final glance at the dead man's back, plowed off into the desert. After all, he was dead.

I've always imagined that when you finally go that one step too far, there's a second of realization, one instant in which you grasp how badly you've fucked up, and it comes clear as crystal that you won't be walking away from this one. And I've always imagined that you'd know, maybe only for a split second, but for certain know, what stupid or otherwise thing you'd done to deserve this.

But halfway to Tuba City I still couldn't figure out why this guy had died. The highway was razor straight for miles in both directions from the spot where he'd kissed the grillwork of the Peterbilt. Traffic on this Monday was next to nothing. And the day . . . the day was one of those Arizona days where the sky spreads blue across forever, where you can see every feather on every buzzard, every hair on every prairie dog, every stick and stone from here to Albuquerque.

He had to have seen that truck. If he was looking at all, he had to have seen it. So how come the Jeep lay crumpled in a ditch on the wrong side of the road, and how come there was a man face down on the asphalt—dead as last Sunday's lasagna in the ditch beside the old road?

Highways have a way of scrubbing your mind, a way of numbing you with their endless tunes and unbroken distances, their pointless curves. But a hundred miles farther on I couldn't get the dead guy out of my head, could still see the blood creeping off the road, down the dry wash, could still not figure what had pushed him into the path of that truck.

I turned off the pavement onto a dirt side road and stopped at a wide spot where you could see all of the nothing that surrounds this part of Arizona and long strips of empty road in both directions. I got out to see if I could walk the picture, the questions, out of my mind. A mile or so out into the creosote and sage I found a wedge of rock big enough to sit on.

It was midafternoon, the temperature was near one hundred. Nothing was moving. The image of the man's shoeless feet and his blood trickling off the greasy roadbed into the mouth of the arroyo came back to me. The May wind worked at my hair, sucked at my breath. I thought of how, with the next rain, his blood would move a little deeper into the desert arroyos. And how, with the thunderstorms of July and August, the iron-red heme would roll with the boiling mud into the gorges of the San Juan, then into the Colorado itself, across Arizona and California. Then, as the storms above the desert thickened, the thin red curl would reach all the way to the Sea of Cortez and rise as a coral-pink tint in one of those sunsets beyond Land's End admired so much by the tourists. And at the very last it would dance as a sanguine glint in the glass eyes of St. Joseph as he stands in the church window there, watching the sea.

The afternoon wrinkled around me. Then I heard it, so faintly at first, I thought Iimagined it. But if I was still, and breathing very softly, a long way off, I could hear the river. I could hear the water, a full thirty miles from where I sat, rolling across slickrock, spouting over fallen boulders. I could hear the silt whispering as it ground past the bedrock, hear it gathering behind the dams and settling on tidal flats full of egrets, and stilts, and herons. And if I was very still, I could hear a thin green surf where the water at last fell upon white sand beaches.

The gray-green desert and metallic blue sky faded, and where they had been there was only the sound of the river, the falls, the great flocks of birds, the pulse of the far-off surf.

I don't know how long I sat and listened—an hour, maybe two. But the whole time an urge to follow the sounds, an urge to see the Colorado, beyond the hills to the west, grew in me. Never mind that the river was miles away. Never mind that there were no roads and no water in between—a wish to see the river, see the river now, blossomed in me like a tumor. Nothing else mattered. I sat with it. The sun moved farther across the sky. Then at the last, I walked off and left it there.

Near the car I turned to where the sound came from—turned to look across the flaming globe mallow, the cactus and the sage, to imagine beyond the junipers—and listened for a moment longer. Maybe he was drunk, really drunk. Maybe he was struck blind by an instant of insight. Or maybe he'd just had life's spit on his face for the very last time. No way to know.

With an effort I closed the door on my truck and started the engine. I grabbed a Jerry Jeff tape, shoved it into the deck, and turned up the volume. Then I just drove off, toward California, like I knew exactly what I was doing.

The Thing Inside Your Head

On the last night we spoke, we sat
as we always did, on cracked wooden chairs
with cups of yellow wine, and as always,

we walked about in the ditches of his life,
weaseled about
in the weedy places of his mind.

But that night, when the words came to the spot
where he'd lost his hand
in the green-toothed jaws of the combine,

the spot where he'd left the land
and hired on to the barges pushing garbage
down the soiled river, when the words finally arrived

there, he didn't cry, as he always had.
Instead, he idly reached in through one socket
and pulled out his skull—

    just the bone of it. For a moment
he simply stared at it, as though surprised
    to have found something as ordinary
as a bone inside his head, as though embarrassed

    for a moment by the banality of it.
Then he set the skull aside
    propped it at the corner of the blue plastic table.
Suddenly, we talked on, both of us,

as though nothing had happened,
    while the bone, thinking no one was watching, collapsed
into a beetle, and skittered off
    the table, across the boards and the bits

of black and white linoleum. Below the sink,
    it stopped, reached with its crooked feelers for a crack,
then with a rattle, flattened
    its black carapace and slipped beyond the chipped baseboard,

below the clogged drain,
the rusted iron wool,
    below the clotted dish soap.

And then we just sat.
I stared some at his wrinkled head. He
picked at his nails and looked about

as though something more
should have happened.
The dishes, the ones with the lavender

stripes, slowly shed their glops of food
into the oily dishwater and a moth
bashed repeatedly into the 60-watt bulb

above the table. I tried,
but could think of nothing
more to say. So

I left him there, earlier than usual
that night, turned out the porch light,
and locked the back door as I always did.

And I walked out, beyond the slow rasp
of his breath, walked out
and left him there
beneath the electric sizzle of the awful stars.

His was a body full of slow bones, after all,
and he moved as if long fish swam in a small place.
— Alberto Alvaro Rios

Death Wish

People leak out of this world in different ways. My children's mother found a hole torn big as her body in the earth's rough fabric, and with the help of a little carbon monoxide pushed herself through all at once. My brother-in-law's mother chose a coarse seam in the same fabric and bled through it a little at a time—bled so slowly that she forgot what she was up to and left her body behind with no one inside. It was several years before her body noticed what had happened and quit trying. My wife's mother sent her breasts first, my grandfather sent his right leg. And my Uncle Harold chose to tear his own hole in the fabric with a table saw, then stepped through to the other side in such a hurry that he left his hands behind. Those hands spent the better part of an hour looking for something to fiddle with there on the rip fence of that saw. Then they gave up and hitched themselves a ride out of this world.

Different ways, lots of them. And that's good, because everybody's got to choose one, sooner or later.

This hardly seems the morning to be wallowing among these black flies. The sun is shining like the day it caught fire, the air is warm—dead warm, and the Colorado River is right below the whorls of my fingertips. Three boats float ahead of me—my brother Michael and his son John in their canoe, and my sister Pat and her husband Steve, each in a small cataraft. I'm piloting something that approximates a kayak. Together on this morning we're floating the last few miles of Glen Canyon—from the dam to Lee's Ferry. And by the purest of coincidences, we have picked the very day God had in mind for a float when he (or she) chiseled with a silver gouge, this stone trough. The water is the color of Cleopatra's emeralds, the sheer walls the red of Saint Stephen's blood, and the sky is snapped just perfectly blue to the iron rivets in the canyon's rim. A day as fine as most any I can recall. But. . .

The last of Glen Canyon, near Page, Arizona

Today I'm thinking about dying. Maybe that's because I know where this river is headed, maybe it's because I'm on the verge of turning fifty, or maybe I just can't face a really fine day without punching a hole or two in it. But. . .

Today I'm seriously thinking about dying.

People have laid claim to one criterion after another as defining the ultimate difference between animals and humans—language, tools, self-perception, on and on. One by one, though, each of these has been eliminated as the quintessential human trait. Chimps use language, ravens use tools, and anyone who's ever watched a dog with a squirrel cannot possibly doubt the dog's or the squirrel's self-perception.

But this thing, this thing I'm thinking about, this thinking, seems peculiarly human. Other living things on this planet don't think about dying, or at least don't seem to. Maybe that's because most living things on this planet don't die—at all, at least not from "natural" causes.

We round a short bend in the canyon and work our way into a shaded spot under an overhang of sandstone approximately the size, weight, and color of the planet Mercury. Beyond that the resemblance fails. It's deep cool in the shade. Great streaks of varnish slick the wall overhead. Ah, if only I spoke stone.

The current presses us close to the wall, dead-vertical here. I drag my hand along the roughened face of the rock—feels like lizard skin, rock lizard, of course. Stones don't die either, though some might argue that stones don't live. Others might argue that stones' fates are worse than death—being slowly split and crushed by wind and water, cryptogams and piñon pines. I don't know. But I don't need to pester my friends the stones to make my argument this morning. Things much more obviously living will do fine.

Success on this planet is a difficult thing to quantify. But from a biological standpoint (a standpoint of intense interest to biologists but of surprisingly little interest to most other living things), species numbers, diversity, biomass, and geographic distribution are pretty good indicators of success (at least Darwinian success, as opposed to Bill Gates–type success). By these criteria the most numerous and diverse species are the most successful species.

And on this planet, that's bacteria. Nothing else even comes close, or has for the past billion years or so. And among bacteria—the most numerous and successful of all living things—there is no such thing as death from "natural" causes, only growth and fragmentation. Nor is there among most of the bacteria's first cousins—the fungi—death as we know it. And it doesn't stop there. For quite some distance up the evolutionary ladder (or across the evolutionary bush, if you, and you should, prefer) things don't die. Even things as complex as trees—bristlecones, redwoods, and especially aspens—don't seem to die from "natural" causes either. There are aspens in this world that have been alive for hundreds of thousands, maybe millions, of years. And today, while the air smells as sweet and dry as new linen, that's pissing me off some.

Steve and his small boat rotate slowly toward me like a wind-spun leaf—a leaf with a border collie and a bearded and browned middle-aged man on board, that is.

 "How goes it?" he asks, his voice ricocheting across the stone wall as he pirouettes past.

"Extraordinarily well," I respond, keeping my thoughts to myself for the time being.

Steve is long and thin, full of river water and anarchy, and bearded a little like Gabby Hayes. Steve is mostly the color of old wood because, as often as is humanly possible, Steve is naked. He hikes wearing nothing but a baseball cap, tennis shoes, and sweat socks, though for the chance encounter he carries a pair of gym shorts stuffed into one of his socks. We've discussed approaching REI and the Swiss Army about manufacturing ankle packs. Nothing formal yet, still in the planning stages. But the idea is ours. I want you to remember that.

"Did you see the heron?" He asks.

"I did." The blue-gray bird stood more than a meter above waterline, the color of morning fog and still as a stick as we floated with ten yards.

"Lovely ploomage. Eh?" he asks, parodying John Cleese, then rotates beyond hearing. Josie, his border collie, peers intently down-, then up-, then downstream again, her head resting solidly on Steve's shoulder.

Abruptly, it's just me and the river again. My kayak seems to be the slowest of all the boats—something about the relationship between speed and hull length, and that leaves me in sole possession of last place with a pancake of solitude for company. And even though people-type company doesn't come any better than that ahead of me, the solo time is fine by me. The river pushes all of us slowly south, each careful with our little craft, like a spider with a fly.

The canyon widens some, low sandy beaches appear, and pygmy juniper, sage, and tamarisk; the sounds soften. A canyon wren preaches to the choir. With the shoreline more distant now there is almost no sense of movement. It feels like crossing a lake very slowly. Unless I look down. There, the moss and the rocks on the river's bottom, clearly visible six or eight feet below, are streaking past like a bearded rock snake.

So if all these lesser life forms don't die unless stepped on by a cow, invaded by infectious crud, or struck by lightning, how come I have to? How come I'm growing old and feeble and will eventually crap out all together? There. I've got it now. I've got what's really eating at me this morning. Why can't I reasonably expect to go on floating sweet green rivers like this one with dear friends like these for as long as I want?

I'm back in the August (august?) sun now, dragging my paddle tip right, then left, watching the emerald-green eddies curl out behind. My sister Patricia passes me, sailing her craft backward toward the Sea of Cortez. She smiles. I imagine Achilles and Agamemnon in their great ships. I smile back at Patricia as she recedes in the distance.

"Anger be now your song, immortal one." Thus the voyage and *The Iliad* begin. Homer had it right, too. But even he, in fact especially he, in the dungeon of his blindness, knew that the song's the easy part.

Rivers don't die, either. At least not like people do. This river has flowed more or less just where it is today for over 2 million years. Two million years! When water first started working its finger into this crevice the Rocky Mountains were nearly new, the last of the great cats had just wandered into the La Brea Tar Pits, and *Zinjanthropus* and *Homo habilis* were chipping knives from bone somewhere near Olduvai Gorge. As long as there have been humans, there has been this river.

Patricia shouts at me to eddy out, and with a small push from the river's large hand I run my kayak up onto a thin strip of beach between the red canoe and the aquamarine catarafts. For a moment we stretch, and each of us admires the river in his or her own way. Then three snout-rigs—tourist boats riding on silver air-filled bananas—slice past, each pushed by twin Evinrude 150s, each making maybe thirty knots dead-head, up-canyon, each with the voice of two dragons. The boat captains wave, we wave back. The sound quickly dies, and then all that's left is the slap of the big boats' wakes crashing against our little bit of shoreline.

John grabs his rod, a couple of spinners, the optimism of a net, and heads off to fish—wading, but waderless. I don't know how he does it. I've been in snowmelt creeks that were warmer than the water is here, just below Glen Canyon Dam, where even the last of the warmth has been squeezed from the water. Well, he's only twenty-one, give him thirty years or so, and see how he likes it then.

Steve stacks the two catarafts and begins to unload the kitchen box, using the rear deck of the upper raft as table and stove and the lower craft as a shelf for spices, pans, what have you. A regular kitchen on blue pontoons. I roll out a sleeping bag and pad, place a couple of 200-million-year-old rocks in each corner for esthetics and for ballast, then dig my wine and folding chair out of the canoe. The four of us sit then to watch the river as though its flow had just begun—this very moment or maybe the one before. Michael uncorks a bottle of tequila. We pass it around.

John, on his thick and now pink-with-the-cold legs, works the green riffles before us. His casts are short and slick. John's a junior at New Mexico State University in Las Cruces, another victim of the sirens of biochemistry. Firm of jaw and bicep, hair the color of sand, he reminds us, without intent, of how long ago each of us was born, how time, in its channel, flows.

We sit for a little, and then bubbles of conversation begin to rise. The wind has risen as well, and our conversation settles on tales of wind. Michael and I recount our last night on the White Rim this past April.

That May night we camped at Labyrinth Canyon— not much of a campsite, but better than Moab. The wind was gusting a little, enough to make pitching the tents a challenge, but not too bad. Like most places on the White Rim, there wasn't enough soil for tent stakes. So I tied each of the tent's stake loops to a big, I mean *big*—five- to ten-pound—rock. Then, for good measure I placed a plastic tub full of clothes—maybe twenty-five pounds, dead center inside the tent. I expected it would be there if I needed it later. I had no intention of sleeping in the tent, of course—I hate tents. This was a precautionary measure, in the event it rained—I hate rain more than sleeping in tents. And the sky this April evening was wrinkled up like the air over a cast-iron skillet. So, with my tent in place and properly weighted I retired with a glass of wine to watch Mike make dinner. By the time we headed for bed, lightning was beginning to tattoo Bighorn Mesa east of us.

Michael and I settled into our sleeping bags and quickly fell into the sleep of the just. Now, among the many things none of us had sufficient knowledge of at that moment was one critical fact. The Labyrinth Campground lies smack in the center of perhaps the greatest, and potentially meanest, vortex in the western United States (excepting Sedona, of course). The mouths of five deep red rock canyons open onto the Labyrinth Campground—Upheaval, Horsethief, Taylor, and the northern and southern sections of Labyrinth Canyon. Five long, deep, in places narrow, air-sucking canyons. A lot of nitrogen and a little oxygen were about to teach us some physical geography.

About an hour after falling asleep I partially awoke and in a dreamlike state noticed my tent seemed considerably closer to me than when I had rolled into my bag. I watched for a minute more, trying to scrape the lint from my brain and make some sense out of this. As I watched, the tent performed a segment of what appeared to be an energetic and complex flamenco. Then it skittered about twenty feet to the south. Now I was awake. However, as will be apparent momentarily, my logical mind, the portion capable (at least occasionally) of rational decisions, was still sunbathing somewhere near Xanadu.

I scrambled out of my sleeping bag and ran for the tent. I caught it, but only briefly. I stood there for a moment looking a little like a reef-bound windsurfer. Then, while I was trying to imagine my next move, an approximately eight-megaton microburst tore the tent from my hand, lifted it—stones, plastic tub full of clothes, and all—about thirty feet in the air, and sent my nylon home cartwheeling far above the hardscrabble. I (and here's the part where my mind proves itself to be mostly cornmeal mush) take off running for the tent. I have no shoes, no socks, no shirt, and no pants on. The terrain is shale terraces—dragon scale interspersed with five-to six-foot drops—from where my tent first imagined itself a parasail, all the way to the river. All of it bristling with woody scrub and malignant beaver-tail cactuses.

The tent returns to earth about a hundred yards south of where I pitched it. I manage to come to earth at about four spots in between. I do, however, finally achieve the tent—which now looks something like a load of clothes hangers than have been through a dry cleaner's mangle along with the remnants of a downed nylon dirigible. I sit on it. By God, if it goes again I'm going with it. I glance down at my hands and legs. I now look somewhat like I imagine a man might look after a nasty fall while trying to waterski behind a Harley Davidson on I–40.

My clothes were spewed from the tent as it cartwheeled through the sky, and now my socks, underwear, pants, and shirts are strewn across the intervening hundred yards of desert shale. The wind, of course, immediately dies.

I carry the remains of what was once my tent back into camp and drive the car onto it. Then, while everyone else is laughing their asses off, I daub Betadine onto my wounds, pull on my long underwear, and get back into my bag—which, were it not for Michael's quick thinking during Mother Nature's little snit, would be drifting in the stratosphere over Phoenix by now. The rest of the night was uneventful.

Pat and Steve are appropriately impressed. They always have enjoyed stories of Michael's and my misadventures. So even though I notice a certain eagerness in my sister's dark brown eyes, I relate one more tent story—one in which I was not a direct participant, thank God.

Two friends, Larry Metz and Bill Marquardt, camped one evening near, as I recall, Westwater Canyon on the Colorado River outside of Grand Junction. After dinner they hiked to the rim of the canyon to watch the river juggling silt among the rocks below. While they were watching, an odd-looking craft came round a bend upstream. It was red and green and tan and looked a little like one of those old bull boats that plied these waters in the 1800s—a little like an overturned washtub, but larger—six, eight feet across. But this craft was closed on top by what appeared to be a layer of green nylon. As it moved closer, Bill said, "You know what that looks like?" Larry asked, "No, what?" Bill replied, "It looks like a tent. In fact, it looks like MY tent!" It was.

Chuckles all around. Then Pat says, "I can top that." And she tells a story of camping one evening in the Grand Canyon when, during an evening walk, someone sighted what appeared to be a UFO (unidentified flying objects are commonly seen in canyon country) circling well above the rim of the inner canyon. Again, the markings were unusual—beige and olive green taffeta—and the motion a little erratic, even for a UFO. In fact, this object was a little tentlike, but it was flying toward Burma, near as anyone could tell. And a tent it was, carried completely above the rim of the inner canyon by another one of Mother's playful gusts. The tent, quickly pitched before an evening walk, belonged to one of their group. Tents were shared for the remainder of that trip.

Everyone agrees, that's tops. The finest tent story we've heard.

While we are topping one another's stories, an odd thing has happened. And now that we're sipping our drinks and our thoughts return to the river, it's dramatically obvious. The river has dropped. I don't mean dropped a little, like somewhere upstream a few brontosauruses are taking a drink. I mean dropped a lot—maybe a yard, and even as we watch it's clearly continuing to fall.

"They aren't allowed to do this," Patricia says. "There are limits on how much they can raise and lower the water."

She's referring, of course, to the folks with the concrete spigot—the Bureau of Reclamation and Glen Canyon Dam.

The water drops another foot. It's hard to watch. It's like being there at the end, when Phoenix or Los Angeles siphons off the last of this world's water. Another foot. Wide stretches of mud and shale are exposed now along both sides of the river, only a center channel flows thick with water, and still it drops.

I think of the trout beneath and wonder what they are thinking. Do they imagine this is the end, or are they wise enough not to waste what little time they have left on useless thoughts of whether this is an ending or a beginning? Another foot.

Parts of the riverbed look like the insides of roadkill, the color of flesh that has been underwater for a long time. We are each of us inside ourselves, awestruck.

"Damn it! They are not allowed to do this!" Pat again.

"We'll make some phone calls when we get back, Pat." Steve this time, sensible.

I hear them, but all that registers with me is the sound and look of the river. It's been turned off. Like you might turn off your garden hose, or the water line to your toilet while you replace a leaking ballcock. Like a faucet whose dripping has annoyed you long enough. Off.

Finally, it stops—maybe five or six feet down from where it was only an hour ago. The 2-million-year-old Colorado, just like that, reduced to a sluggish trickle in the trough of a five-hundred-foot-deep canyon. I'm stunned.

"Does this go on all the time?" I ask Pat and Steve, because they've logged months on this river.

"No, it doesn't. And it isn't supposed to. I've never seen anything like it," Pat tells me.

We look as long as we can. Then we turn our backs to the river and make dinner. I clean and poach two trout—rainbows John has pulled from the shallow water. I place a little of the trout on each person's plate. We eat the delicate white flesh, and I think of the fish in the river's shallows, desperate. I think again of Homer and *The Iliad,*, and of the priest Kryses, pleading for the gods' help while Agamemnon holds his fair daughter:

> if to your liking
> ever in any grove I roofed a shrine
> or burnt thighbones in fat upon your altar—
> bullock or goat flesh—let my wish come true.

The wine helps. The tequila helps. The trout help. Homer helps. But it isn't nearly enough. So we simply ignore—as the light fades—the river in its bed.

John rolls his pad and bag out on the new wide beach. "Maybe that's not such a great idea," Michael says. "The river may come back up in the night. Not a great way to be awakened, river inside your bag."

 John, and the rest of us—most of whom were considering the same thing—move to higher ground. A little conversation follows, then people drift into their own thoughts.

There's a kind of death that is going on inside all of us, all of the time. Even before birth, parts of us are dying. It's called *apoptosis,* this death within us. Apoptosis comes from the Greek word for dropping—like petals from flowers, like scales from snakes. It's in our genes, God or Charles Darwin put it there. Every day, part of us must die. If it fails to die, life itself will kill us. Life—unrestrained—is so dangerous that things like cancer and autoimmunity are all triggered if we aren't dying fast enough.

Some of our cells, maybe most of them, are even born dying. And these dying cells live only so long as neighboring cells "save" them. We get by with a little help from our friends. When the help stops, cells start to die. For them, then—these cells born dying—life isn't so much living, it's simply another moment of not dying. Each day is a reprieve; there are no givens. Each day another temporary stay is handed down by some stranger. Even the lenses of our eyes are made from the carcasses of cells that died years ago, died when there was no more help. Everything we see is filtered by those dead cells.

There is a monster at the very middle of us devouring a part of each of us each day. It seems little wonder, then (doesn't it?), that someone like me, who is so filled up with, has always been filled up with, dying is unavoidably drawn to death—its words, its appearance, its inevitability. But even the iron of inevitability doesn't dull the teeth of fear—not ever.

I sleep fitfully.

Morning. And "dawn in her saffron robes comes spreading light on all the world."

Strong words for a blind man.

I stir from my bag. The canyon is still as deep and red as it ever was, the sky is still stapled to the rafters of heaven, and the air is heavy with water. Eggs and bacon are frying somewhere near, and my stomach recommends a commitment to one more day. I rise. And to my great joy, my first vision is of the river at its former depth. But I cannot shake the image of the concrete spigot, just up-canyon, and the hand of some angry god with the face of Floyd Dominy raising and lowering the river for his amusement. (Later, I will learn that this strangulation was a singularity, created by a major power outage in the Pacific Northwest.

Washington and Oregon, parts of Idaho and Montana, for no reason anyone can imagine, went completely off the grid. No place to ship the fearsome power of Floyd's great turbines. Shut off the water! Turn off the fucking river, now! That discovery, of course, will only serve to etch the spigot image deeper into the abused folds of my brain.)

Nevertheless, just now each of us is buoyed by the river's reappearance, even the river.

The trip out is glorious. Long green stretches of solitude, trailing a hand or a foot in the cold water. Nosefuls of water and sage. Another heron and then another. A golden eagle.

The author exploring a tributary of Glen Canyon

I don't know if apoptosis has anything to do with how our bodies finally die. But our deaths are like the deaths of our cells in one respect at least—they're programmed into us from the start. Death is in our genes, and it's there for a reason. I, of course, have absolutely no idea what that reason is. But if bacteria and trees don't die from natural causes, then our deaths are no accident. Death is a gift. Six hundred million years ago or so, life coalesced in our seas. Somewhere in the intervening millennia death arose to repair life's major flaw—the oversight of immortality.

Michael and John in their canoe are next to me now. John has caught two more fish. He's obviously very pleased with himself.

"Uncle Gerry. Can I try the kayak for a bit?"

"Sure."

I transfer my camera and assorted paraphernalia to the canoe. John steps into the kayak. I explain a little about paddling to him and he's off.

"Front or back?" Michael asks.

"I like to drive," I respond.

"Back it is then," Michael says, and we arrange ourselves rather ungracefully in the canoe. I push off with my paddle and once again we're in the bosom of the river. We talk some, but mostly we just float, often spinning slowly across the river's wide palm. The day is wrapping up its few remaining tasks and beginning to look to the west. Michael's a physicist by trade, a human being and my brother by birth, my friend by choice.

"Michael," I say, "I've begun to think that much of what we imagine to be reality is created inside our own minds." He surprises me and agrees. He tells me a story of looking for ibuprofen one morning under the sink and having it appear only when his wife told him that what he was looking at was in fact what he was looking for—it wasn't there until he expected it to be there.

"At times, I wonder if any of what we see is really 'out there.' " He muses.

"Me too. At other times, I wonder the opposite: What's become of what we know should be out there?"

I explain my pet theory about how things sometimes slip forward in time. When you look for them they aren't there—yet. Wait a moment or two, then look in the same spot where you just looked, and presto, they've reappeared. I call it my temporal lobe theory of virtual reality and special geography. Nobody much gives a shit.

That night at Pat and Steve's place I volunteer to make dinner. Devil's Backbone Burritos, of course. While I'm pulling it together, Mike and Pat and Steve watch a video about people who've run the Colorado River in either direction. John reads, and the stars appear from wherever they just were, softly shaking the night's bell.

I'm slicing onions and garlic. And again, I'm thinking about dying.

So if that's true, if death came to us as a gift, as a talisman to guard us from the false finger of life, why do we worry about it so? Why does it hurt so damn much? Did we learn that, or is there some utility in grieving over our own demise, something useful about the pain of losing a friend?

On screen some idiot is trying to push a jet boat upstream against the foaming fist of Lava Falls. Again and again the boat is slapped back. Again and again, the pilot pushes the boat back into the fierce rapids.

They dance like that every day for weeks. Until one day, for no imaginable reason, the man is allowed to pass. In the films he looks oddly disappointed. Maybe I give our species too much credit. Maybe death is our only salvation.

I pour another glass of wine, and while finishing the onions and trying to watch the TV, I cut my finger. Not badly, but deep enough to bleed freely. The tip of Pat's knife is broken (probably from bear hunting) and I wasn't paying attention. I look at the opening in my flesh. I watch the drops form on my fingertip. Deep red, the color of old stone.

As I reach for a paper towel to stem the flow, a drop of my blood falls into the bubbling pot of pinto beans. I stop. Another drop, then another falls into the pot. I watch my blood deepen in color, then disappear into the soup between the beans. I look at those watching the TV, the stars beyond the open door, the dog, the lights of the Navajo Power Plant and Page, a car passing on the highway. I spoon the frijoles into their tortillas, roll the tortillas, and place a burrito on each plate. I don't mention the blood.

When they had finished
raising the barrow, they returned to Ilion,
where they all sat down to banquet in his honor
in the hall of Priam King. And so they performed
the funeral rights of Hektor, tamer of horses.

Unemployed in Greenland

A man is sitting in a poorly lit tavern in Godthaab,
on the south coast of Greenland, making
idle doodles in the spilt beer at the bar.
Another man is dancing across the scuffed
wooden floor, holding his arms as though

he was dancing with a partner. The
steps he executes appear rather formal,
like he is dancing one of those intricate ballroom
sorts of dances. Step, glide, . . . step—rotate—
glide, step. No music is playing. The bartender,

a large man with grizzled jowls and an apron soiled
by pickled herring, is reading
a newspaper. We can't tell what the paper says,
it's written in a foreign language. But on the cover, there is
a picture of a once dark-haired woman in a heavy woolen sweater

smiling broadly and holding two large dogs. The dogs
are licking her face. There is a window to the left of the bar
and through that window, we can see daylight, but
that's no reliable indication of the time of day here,
so very far north. All is serene. Except,

periodically, and in association with no obvious
event, the man at the bar slams his head violently
into the polished pine of the bar top
making a sound like the sweet-spot of a Louisville Slugger
connecting cleanly with a 5-pound butternut squash. Each time

he does this, a series of tones rise from beneath the bar
like crystalline sleet blown against glass cabbages.
When this happens, the dancing man pauses, smiles
as though reconsidering some intricate sweetness, and
then resumes—step, glide, . . . rotate. The bartender

ignores them both, continues devouring words
only he seems to understand, words that
might have to do with sled dogs or ice hockey, human remains
in a Dumpster in Stockholm, or a woman killed
for her leather purse and her wooden shoes near Copenhagen. Step,

glide, . . . rotate, glide, . . . thwack! Crystalline sleet. Pause.
    Step, glide, . . .
Beyond the windows, hogbacks of crushed ice roll north, one
after another, for as far as we can see into the blue distance.
The crest of each frozen wave blackened by soot and blown soil.

And over it all, there is the smell of raw fish and hot oil,
the smell of barren stone and the fertile fist of the sea And over
that, a broth of warm lead stirred by a stale breeze. Step,
Glide...—crystalline sleet.

The first step, the crone
who scried the crystal said,
shall be to lose the way.
—Galway Kinnell

Aravaipa Canyon

Sometimes it is my deepest wish to be anonymous, emptied of words and names—sweeping blow-sand from wooden steps at the back of a two-room adobe, goats mucking about in the shrubbery, a pig or two, and waves of wet creosote rippling off in all directions. Overhead, there would be clouds burrowing blue at the bottom of a nameless sea, and in the distance paloverdes dancing like fistfuls of snakes in the sun-wrinkled air. Sometimes I wish that—sometimes I wish it very much.

Water, sprinkled with light-brown foam, laps at the bottom of my nose, gathers in the gray strands of my beard. When you're at eye level with water striders, their feat is even more dramatic. They skate, buoyed only by the few calories that form the glue of hydrogen bonds, and scatter like buckshot before my hands and breath. Standing on dimples of water without a net, without a thought for the depths and the carnivores below. Fish nibble at my genitals and buttocks, and the rest of Arizona politely looks the other way this August day.

I lower myself until my head sinks beneath the surface, and I open my eyes—pale green cream shafted yellow with sunlight, fish big as thumbs staring back at me. Loach minnow, roundtail chub, spike dace, and gila mudsuckers make a living here, but I've no idea which these are, especially viewed through this murky lens. Bits of wood and grass slowly tumble through the shadows, rid of the burden of gravity. Then the buoyancy of my own flesh (one too many burritos last night, I suppose) drags me back to the surface. I rise, eyes only—like Rambo—and reconnoiter.

Watercress, green and thick as Rapunsel's pubic hair, undulates along both banks of the stream. Beyond, willows are palming the wind, and beyond that the wolfish fist of the Sonoran Desert—saguaros, fishhook cactuses, snakes, gila monsters, scorpions.

Aravaipa Canyon, near Safford, Arizona

A vermilion flycatcher, a bird like a crimson meteor, abruptly imagines I'm something I'm not, realizes its error only at the last moment, and splits the air just inches from my face. I take the cue, and pull my body from the water. Naked except for hat, tennis shoes, and sweat socks, I settle my aging flesh among the watercress and warm mud. The temperature is somewhere above one hundred degrees, thunderclouds are gathering like pig iron in the southwest, and just now I'm lost. Never happens when it's convenient, does it? And, as always, for reasons I can't quite lay my hands on, I'm kind of anxious to get unlost. But just now, it'll have to wait.

I drove from Safford, Arizona, this morning, about sixty miles, from Bisbee the day before (left my wife with relatives there), Las Cruces and Albuquerque and Fort Collins the days before that. Left behind the university, the city, my classes, my research, and drove over a thousand miles to get here—correction, drove over a thousand miles to get here and get lost. In spite of it all, though, at this naked moment I'm really pleased to be here, I mean really pleased! Some famous, or should-be-famous, man once said, "What do you mean I can't do this? I'm in the midst of a mediocre career, I can do anything I want!" I know just what he meant, here, ass-deep in Aravaipa Creek.

A long arm of the Colorado River begins around here—water gathers from as far east as Diamond Peak in the Black Range of New Mexico (where the drainages of the Gila and the Rio Grande divide), north from the Mogollon Rim (where the Colorado Plateau begins its long climb to the Kaibab uplift and the waters of the Little Colorado and the Salt Rivers divide), and south from Ojo de Agua in Sonoran Mexico (where the waters of the San Pedro River and the Rio Sonora separate). Ojo de Agua—good name. All that water finishes where the brown swirl of the Gila River collides with the Colorado 300 miles south of here, near Yuma. Well, at least the water Phoenix and Tucson don't take, in wet years anyway, finishes 300 miles south of here.

The Gila is a genuinely magical river. (How else would you explain all this water in the midst of all this drought? How else would you explain its sudden appearances and disappearances?) And the San Pedro River, with its tributary, Aravaipa Creek, are among the most magical fingers of the Gila. Surface water, clear as Waterford crystal on Dublin lace, flowing perennially in the midst of the eternal furnace of the Sonoran Desert—watercress and blue herons hard by vultures and giant saguaro begging forgiveness for their prickly sins, muskrat and rattler, coyote and mudsucker, deer and scorpion. Of places where I might get lost, and there are many, believe me, this one is a hell of a lot better than most. But lost is still lost.

I've come to that point in my life where I can no longer recall the reasons I began most things, the point where all directions look the same and every choice seems equally senseless. I drink too much, I play with myself too often, I eat like myocardial infarctions and colon cancer don't kill middle-aged men, and every morning I swear I'll change, but nothing ever changes. Except times in the desert, and especially except times in the desert near water.

For some it's the ball-peen hammer of smack or the lightning-tight twist of crank. With others it's the rush of a faceful of food, sexual dominance, or dancing with the dry hand of death across some delicate spit of rock. For me it's these split and gnarled drainages, these arroyos, these roadrunners, rattlers, gila monsters, scorpions, whiptails, hoodoo canyons, and the baleful face of the evil moon over places where the waters of the Colorado River gather.

About four years ago I began to notice that in some ways, often important ways, things—breathing for example—made more sense to me when I was waist-deep in some pothole or partway up a slickrock slab in a canyon carved by rain and ice fighting their way to the Sea of Cortez; made more sense in the company of ocotillo than of academics, more sense amid the aroma of wet creosote than beneath the smell of ether and acetic acid, more sense by moonlight than by quartz halogen. So I come here when I can, and I try to imagine other ways the pieces of my life might fit together, and reasons that I could use to go on doing the things I do for money. Sometimes it helps, sometimes it doesn't. But even when it feels wrong, even when the visions don't come, there is within me one who feels nurtured by all of this, one who may yet be of great importance to me.

So I come again and again, hoping that if I touch the desert and the river often enough, some of it will stay within me, will come with me into my classrooms, or sit on the bench next to my notebooks and microscopes, or lead me out of all this and teach me things I need to know, need to know very soon. My wife accepts this, accepts the dents in our checkbooks and the damage to our cars. And she accepts the solitude. She knows, as I do, that when the spirit appears, it appears to me alone, for it is my spirit. She must look elsewhere for hers. In return, I bring her rocks and feathers and bits of wind-polished wood.

I slide back into the water to rinse the rich mud from my naked backside. Twice more I lower my whole body into the water, twice more I settle until the creek closes over my balding head, fills my ears and nose. Twice more, and then I grant myself absolution from all previous sins and the sins of my father, my grandfathers, and those before them. And now, I rise, reborn, and allow the sun's light to fall upon all the parts of my body, even the parts pale from lack of sun and covered by an odd hair, even the parts collecting the creases of age. And for a moment I imagine myself something other than I am, and I imagine that things truly will be different, truly, soon.

How, you might ask, does one manage to get lost in a river canyon such as the one carved by Aravaipa Creek? It came to me as somewhat of a shock as well. But it is clear at the moment that it is altogether possible. Just upstream from where I stand, a fence—bailing wire and willow posts—spans the creek. I walked past no such fence on my way downstream! There are only two possibilities, then. Either a) I have walked past where I entered the creek (the sides are so choked with cottonwoods and low scrub that I could easily have missed where I left my car), or b) the creek forks, and I have walked up the opposite fork from the one I came down.

I did not spend all of that time in school for nothing. So I will use the power of logic and reason to solve this problem (translation: I left the fucking map in the car). This is the desert, right? Deserts are called deserts because they are real dry, have lots of cactus and stinging insects, and have almost no water, right? Okay, that means that it is damned improbable that this creek would even be here, and even more damned improbable that it would have two wet flowing forks of nearly equal size, right? Ergo, I must have walked past where I entered the creek to begin with, right? Impressive this logic thing, eh? Though somewhere in my mind something is clamoring for equal time, I ignore it and begin walking back downstream.

The growth at the sides of the creek is so dense that I must walk in the water. But even if I didn't have to, I would. The water is clear and cool, full of fish, and its gravel bottom offers relatively sure footing. So I walk in its wetness and imagine the full length of this water and the cycles that carry it from Arizona to ocean, ocean to Arizona, Arizona to ocean. . .

A dead muskrat floats in an eddy at the side of the creek. I lift its bloated carcass from the water, smooth the wet fur and lay the body in the young grass at the creek's edge. I place a few green blades of grass on top, and overhead three vultures appear from nowhere. Another set of cycles loops its way through my head and I move off.

I think it was John Wesley Powell who said that even if you took all of the water from all of the rivers of the American Southwest, you couldn't support much more than an average sized rugby team. And though it's pretty likely that John Wesley never imagined the horror of Lake Powell, he was, I'm sure, generally correct. So today I'm particularly grateful to have my own creek and its accompanying canyon and all the water I could drink even in my wildest and wettest dreams. And today the water and the dreams are mine alone!

I keep looking for some egress that I missed on my way up, some opening in the canyon hidden by the trees and bushes that line the creek. Nothing. The defile deepens, and a canyon wren pours its mournful tune down the opposite wall of the canyon. I have walked farther than I should. My confidence in the clearly flawless logic I used earlier begins to crumble. I start looking for something familiar, something—a rock like a face or a tree twisted to just the right angle—something that would tell me I have come this way. Nothing.

Another view of Aravaipa Canyon

Then I notice a snag twenty yards ahead, a log or branch wedged into the rocks dead in the middle of the creek. That was clearly not here when I passed this way before, going up- or downstream. Now it seems that in either direction from where I stand there are things that don't fit, don't nearly fit with what my mind tells me is true! My confidence dissolves completely. Where the fuck am I?

Because I cannot think of anything else to do, I think of Don Juan Mateus, the Yaqui sorcerer, and how he tried to show his apprentice

 Carlos Castaneda the art of "not doing." Unknown to Carlos, Juan had hung a piece of colored cloth from a bush nearby. When at dusk Juan told Carlos to look at the mountains through the branches of the bush, it appeared to Carlos that an enormous colored cloud or membrane was writhing in the sky over the distant mountains seen through the branches of the bush. Then Carlos's eyes changed focus, and he saw the strip of cloth worried by the wind. Pleased, Carlos turned to Juan and told him that he had discovered the trick. Don Juan said disappointedly to Carlos, "For a moment, you nearly had it." Or something like that: "For a moment, you nearly stopped the world."

For me, Don Juan was trying to show Carlos how much of what we see is there only because we anticipate that it will be, that much of our reality is created by our expectations, and if we could rid ourselves of our preconceptions, the world itself would change. We regularly miss opportunities to truly see things because we refuse to accept what we see for what it is. Instead, we make what we see into something familiar.

So, again because I can think of nothing else to do just now, I sit and stare at the wooden snag, allow my eyes to unfocus, and look for other shapes, other things in the wood. With little effort I find the shape of a deer. The image solidifies, and though it looks a little small, I think, "There, I've made this broken limb into a miniature deer." That pleases me. But another part of my mind reminds me that I am still lost. I stare at the snag for a moment more before my thoughts must uncoil themselves and deal with the paradox this snag has created.

Slowly, so slowly that at first I think I am imagining it, the snag begins to move. More quickly now, it turns sideways and acquires the full shape of a young doe. The deer stares at me. Every thought gone, I stare back into her black bottomless eyes. The doe turns then, steps quickly through the water, and vanishes behind the low cover at the creek's edge.

The snag gone, the creek coalesces into something familiar. I've been here before, going down- and up-canyon. Without thought, I turn and begin walking back upstream. When I come again to the confusion of the fence, I follow it south out of the water, away from the canyon wall. In less than fifty feet I walk out from the dense undergrowth near the creek and I see my truck, just where I parked it.

I walk to the rear, drop the tailgate, and take off my wet shoes and socks. I put on my sandals and have a long drink of water, and I just sit for a while. Canyon wrens and another bird or two I don't recognize mark the moist air with their songs. The wind works the willows overhead. Then I reach for the map I left behind.

The creek forks. So much for higher logic. Two branches, nearly equal in size, separate just downstream from where I'm parked. When I walked in, I headed southwest from the car and without realizing it waded into Turkey Creek just above its confluence with Aravaipa Creek. On the way out, again without realizing it, at the confluence I walked up Aravaipa Creek. Walking in either direction, I had failed to notice that there were two creeks, possibly because of the heavy undergrowth along the banks and the frequent channeling of the water—or possibly, simply because I couldn't imagine two creeks in this August fire. But it's clear, the whole time I thought myself lost I was within fifty yards, maybe fifty feet, of my car. Next time maybe I'll remember the map. But probably not.

I sit for a while longer, enjoying long drafts of the sweet air. What a remarkable place. An island of birds and mammals and fiercely green plants isolated by the molten Sonoran sea. As I said, not the largest, but surely the most magical finger of the Gila, the last great limb of the Colorado. And as I said, for me it is deserts and especially the wrinkles and gorges, the cholla and the ravens, the mesquite and the mistletoe, and one deep-eyed doe that wanders in places where rivers gather.

I close up the truck, start the engine, push the lever into low-range for the stream crossings, and start back. A pyrrhuloxia, gray crest and heavy beak, watches from its perch as I close the barbed-wire gate behind me. At the first stream crossing the truck wallows into a pothole. And I stop for a last taste of this place. Again, a vermilion flycatcher mistakes me for something I am not and like a scarlet dart slices past the car inches from my open window. An island of birds.

I shove my truck back into gear, and together we lumber up out of the creek. I drive, too fast as always, down the dirt road. And with a plume of dust rising behind, I imagine myself at the helm of the shuttle as the primary fuel tank turns liquid oxygen and hydrazine into water vapor—imagine myself escaping the unrequested burden of this planet's gravity.

Where the dirt road meets the pavement I slow the shuttle, then bring it to a full stop. I choose not to think about where I've just been or where I'm headed, or even about an adobe home with a couple of pigs and a chunk of land reeking of wet sage and creosote. Instead, I look carefully in both directions, wait for an old Ford pickup to pass, then pull my truck out onto the pavement and drive off toward Bisbee.

Autobiography

It would be a movie, I
think. Black-and-white. It
would star an unshaven Walter Matthau, or
an aging, overweight, Marlon Brando.
It would be difficult to watch, even—
perhaps especially—if you knew me.
It would open in a
small town—two sun-bleached gasoline pumps—
glass balls revolving in their heads;
a market with a single side of beef hanging
in the cooler; and a two-story, wood-frame
house, whose rooms have a certain "dreamlike" quality—
in northern Colorado or southeastern
Kansas or anywhere
in Utah.

Near the midpoint of the film, there would be a scene
with Marlon smoking a cigar and watching a couple
on the beach while he crushed kelp pods beneath
his sandaled feet. His gaze would seem unfocused,
but you'd know for certain he was watching
the two of them. Beyond,
an angry surf is stripping a winter beach and curlews
are plucking bits of shrimp and crab from the thinning sand.
Then the camera, as the couple's caresses become
intimate, closes on a desert scene and an eerie line
of dust devils carving their way between paloverdes
near a pool of tepid water. A coyote would be drinking
from that water. A boy would be watching him.

Then, near the end of the film, Walter,
with his great, gray dog of a face, would have one too many
vodka gimlets and while trying to tie his shoelaces, he would
topple, smiling, from the end of a wooden pier.

He would drown of course—as we've known he would
from the very start—in a few feet of salt
water, his shoes still untied.

"THE END," in eighteen-point Gothic
would appear on the screen. Then the credits would roll—
director, producer, Marlon's name, Walter's name, the list
of supporting cast, best boy, first grip, special effects—ratcheting
past. Finally, the audience would rise, push their popcorn buckets
beneath their seats, and head for the lobby. As they waited for their wives
and husbands to return from the restrooms, there would be comments
like "They could have done a lot more with that." Or "I liked
the part where Brando ate that raw egg whole." Or "I've never
understood anything that director did."

While they speak, an usher
enters the theater from the rear, gathers the soda cups and oily
cardboard tubs, sweeps the grit from under the seats, and turns off
the overhead lights. Then as everyone is nosing their Buicks
out of the neatly striped parking lot, a final scene flickers across
the glass-embedded screen—a scene in which Walter has a last
amphibious dream. In his dream, Walter is young, walking a long
strip of beach. The waves breaking are that iridescent green
only truly seen in paintings. Walter looks strong. He is walking
the beach with a beautiful young woman. As they walk, Walter
caresses the young woman buttocks. "Holleey Molleey!" Walter
seems to say, and then smiles that unshaven lop-eared smile
of his. The dream dissolves. The final shot is a 14mm,
aspherical close-up of Walter, dead in spite of his dream,
floating, face down, in the midst of a school of mackerel, his shoes
mysteriously tied now. His face still creased by a smile. The screen
goes abruptly black, and the movie really ends.

There is a wolf in me . . . fangs pointed for tearing gashes . . . a red tongue for raw meat . . . and the hot lapping of blood—I keep the wolf because the wilderness gave it to me and the wilderness will not let it go.

—Carl Sandburg

Yellow-Crowned Night Heron

He's cheating. I can't see how, but I'm certain he's cheating. A glance at the scorecard and his frustration is clear. As the game has progressed, the column headings changed. At first they read "Gerry/Les," then after a first loss, they changed to "Butthead/Les," and now they read "Dickbreath/Les." Les, nearly forty-three, the father of a five-year-old girl and hard into a career as an industrial biochemist, is not a basically dishonest man. But Les is a fisherman, I mean a *fisherman,* and two days of rain, especially two days of rain on the Sea of Cortez, will change that sort of man—usually for the worse. Gin rummy was all that was left to us just now, and as I said, just now Les is cheating.

Les Walker, the card cheat (a clean-shaven incarnation), on the road to Mulege

Three days before, we had driven from San Diego to Mulege, a sort of semiprecious-stone of a town halfway down the Baja Peninsula. We were staying in a bungalow at the Hotel Serenidad on the Rio Mulege—the only navigable river in all of Baja California. And though you probably couldn't navigate very far, there is a forest of date palms, enormous cardon cactuses, and, of course, lots of good cheap beer.

I had come for the beer and to reflect on the vast flats born from the silt of the Rio Colorado. Les had come to fish. Neither of us had come for the gin rummy. And I still couldn't figure out just how he was manipulating the cards. But he was.

Two days ago we had gone, well before dawn, to the weathered docks at the mouth of the Mulege and hired a boat. One Señor Escobar—a middle-aged man with a sunset over water in his eyes and skin deeply creased from many days and nights at sea—rented us a boat. Then, *then,* he mentioned that there were no bait fish to be found. "But," he assured us, "lots of yellowtail are being taken on candy-bar jigs." This seemed a bit unlikely to me, because every yellowtail I ever caught was taken on the surface, but my thoughts were still murky with sleep, so I simply moved off toward the skiff at the end of the dock. Les, though, had planned even for this unforeseen complication. Men who love to fish and don't get away from work very often are like that. Among the various bags and boxes Les was carrying was a particularly nasty-looking little cooler full of three-day-old mackerel. No bait fish indeed! We'd kill 'em. Just in case, though, Les loaded all the candy-bar jigs he could dig from the bottom of his tackle box, then moved to the back of the boat to sharpen their rusted hooks.

A boy, the boat-owner's son, would pilot. Hugo (pronounced "oogo") was brown and tight as the night itself. Maybe sixteen, with a smile like the moon's. I liked him, in his white T-shirt, yellow rubber boots, and faded jeans. I liked his even teeth and crow-black hair. I liked the way he moved from boat to dock beneath the light of the bare bulb suspended from the pier. And I liked his laughter and the silver in his eyes.

We loaded our gear, and Hugo moved quickly to help with the rods and reels and tackle boxes, the camera bags and the coolers. I stepped into the rear of the boat, next to Les, and straddled one of the bench seats. Dark water slapped rhythmically at the lapstraked sides of the skiff.

Hugo gave a quick jerk to the nylon rope tied to the flywheel on the outboard, and the motor came quickly to life. I cast off the stern line, and we moved down the estuary toward the sea. The lights of the dock dimmed, then vanished as we rounded the promontory at the mouth of the river and moved onto the open ocean. There is no darkness anywhere on earth quite like the blue-black dark above a January sea on a clouded night.

We motored for an hour or so, stopping only once while Hugo peed off the bow. As he lifted seawater to wash the last of his urine from the hull, Hugo asked me to pilot while he finished preparing the gear. He pointed at a beacon on a distant point. I set the tiller and made for the speck of light in the midst of all that darkness. While we ran for the point, Hugo and I made little attempts at conversation—I in distorted Spanish, and he in clear but limited English. We quickly found how far we could go with words but decided it was far enough for us to work together as fishermen. Hugo finished his tasks in silence.

Les and I talked for a while about other fishing trips, colder mornings, deeper waters, the way fishermen do. Then we just watched, Les, sucking a bilious black cigar through his bearded mouth, and I, trying to inhale all of the sea air I could as the black water rolled beneath the blue-and-white bow. The little Evinrude ground away at the sea, and as the water passed from silken to a hard chop, a rumor of sunlight whispered deep blue on the eastern horizon.

Ten minutes later we were at the shoals off Punta Concepcion, thirty minutes or so to full sunup. Les handed me a ripe mackerel and I baited my 9/0 steel hook with the ten-inch fish (dead and then some, of course). We began a slow troll southward as the sun boiled up from the deep green water.

Ah, sunrise on the Sea of Cortez, the stench of three-day-old fish on my hands, ropes of foul smoke curling from Les's cheap cigar, and four cold beers bobbing among the scales in the cooler. Les's gray eyes glint beneath his old San Diego Padre's cap as sunlight skips across the water, then suddenly he pulls hard on his six-foot stout-backed rod as though something has hit the bait. He jerks again and reels hard on the 6/0 Penn Senator. Nope, not this time. But surely it's only a matter of time. Does it really *ever* get any better?

We quickly lost our baits. Some to fish—maybe—but mostly the old mackerel just didn't hold very well on the hooks. When we bought the mackerel at a Chinese deli in San Diego, twelve fish seemed like more than enough. The Asian gentleman behind the counter seemed impressed that we wanted twelve mackerel. So it hardly seemed decent to ask for more. But now twelve fish looked like nothing more than chum. We cut the remaining mackerel into small sections and began to troll these smelly bits in place of whole fish.

Still, and not very slowly, we watched our bait slip into the sea. Twelve fish overboard, no fish boated. The sun had risen, then quickly disappeared behind a low scud. The antihistamines I'd taken before dawn were coming on strong now, and the slow pitch and roll of the sea was pulling me toward sleep. I fought to keep my eyes open, to loosen the knot in my stomach. Finally Hugo decided on a new location, and the wind, as we motored at speed, revived me some.

Due east of the improbable peak of Cerro Colorado, Hugo cut the engine, and the boat's wake pushed the stern a couple of feet into the moist air, then the little skiff settled into place. The sea rolled leaden in every direction, and we began to jig for bottom fish with Les's old candy-bar lures. Jigging for bottom fish while on the Sea of Cortez is to blue-water fishermen what shooting prairie dogs in Alaska would be to grizzly bear hunters—better than nothing, maybe, but damned disappointing and more than a little embarrassing. Les had pulled his ball cap down to his eyebrows and was muttering almost continuously now.

Our disappointment with this day and the absence of fish finally washed over Hugo as well. He dropped his fishing rod and reached into the little oil-stained paper bag he'd carried aboard before dawn. He pulled out what might have been a sandwich, and offered it to me. It was cold frijoles and some kind of fish on a wrinkled brown roll. Because of something the Catholic Church did to me a long time ago, I accepted the sandwich and took a bite from it. To my surprise, it was delicious. I asked what kind of fish was in the sandwich, or something a little like that. Hugo struggled for the English word, and then held his hand edgewise between his eyes.

"Hatchet fish!" I said, not knowing why.

Hugo shook his head. He moved his hand farther toward the side of his head. Now both of his eyes were beyond his hand, and he tilted his head to one side. I looked at him for a long time. Hugo's patience held. Finally, I realized he was trying to tell me that the meat was from a fish with both eyes on one side of its head.

"Halibut!" I said at last, pleased with myself.

Hugo knew the English word and nodded rapidly, pleased with himself as well.

"¡Si, ahleeboot!"

We smiled some more. Then I offered the rest of the sandwich back to him. I knew he had brought little food and I hadn't even considered eating the entire sandwich. Hugo, however, insisted I finish it. It was clearly important to him that I have the food. I gave in then and ate the rest of the sandwich while Hugo had an apple and a banana. I offered Hugo some of the things we had brought for lunch. Politely, Hugo ate a little of each, but not much of anything. At the back of the boat Les continued jigging and cursing, his face nearly obscured now by his cap, cigar smoke billowing from beneath the bill.

It is entirely possible that Hugo's reluctance to eat the things we'd brought for lunch was not politeness on his part but a result of the poor taste Les and I had shown in choosing from the local butchers and bakers (Les and I are known for that. We once bought an ancient—no freshness dating here—package of wild-boar sausage and washed it down with mescal and three-day-old cream-filled sweet rolls while the locals looked on in horror). But this time I think it was just that Hugo was as disappointed as we were about the lousy fishing and hoped that somehow his sandwich would make up for a little bit of this day. The sandwich was good. But the wide gray clouds just kept on covering every patch of the Mexican sky, and the fish just kept on being somewhere else, somewhere else where, we imagined, they were singing lewd songs about our sisters.

Finally, near noon, Les and I decided we could take no more of this—far and away our worst day fishing ever, and I mean *ever.* We told Hugo to call it quits, to take us home. Hugo raised his eyebrows and shoulders in a gesture of hopelessness and apology. Les, a fisherman to the very last, made a final cast, then reeled in his empty jig. Hugo fired up the little outboard, and we headed for the dock. As we did, it began to rain.

As I said, that was two days ago. It's still raining. Because nothing around here is paved, the world has turned to mud, a thick yellowish mud that pulls at our shoes like barefoot beggars and hangs in our clothes like bird shit. The sky is still a gray porridge, except for now and then when the fog thickens and the rain increases—then you can't see the sky at all. We are playing gin rummy and Les is cheating.

I've known Les for more than twenty years; we've fished together, worked together, raised kids together, drunk together, even got laid together (well, sort of). More than once Les has pulled my ass out of a jam, and more than once I was too drunk to deserve it—like that time in the strip joint on the dark side of D.C. I've helped him once or twice, too. So most of the time we overlook the flaws in each other. And most of the time I'd trust Les with anything—my wife, my children, my boron fishing rod (if I had one). But this rain is eating on me, and I am just about to call Les on his cheating when I notice a strange white blob moving across the cracked red tiles in our floor. It seems to ooze from tile to tile, moving slowly toward my feet. I watch with growing horror for perhaps a full minute before I realize that the shapeless glob slowing crawling among the tiles is actually sunlight, fallen from the window, and moving, as the clouds move, across the floor.

I step from the card table toward the window. Les twitches and eyes me warily, knowing even I have my limits. But I look out, and overhead the clouds are breaking up. Ragged chunks of blue sky, like forgiveness, are appearing everywhere. Then Les too notices the clearing skies, and he immediately reaches for his fishing rod and the last wretched bits of mackerel he saved in the truck for just this eventuality. On his way out, he asks me to join him. I pass on the angling. I'm a little off fishing just now. But I can't resist the fair yellow face of the sun.

Together we walk round the bungalows, mud sucking at our shoes, the last of the rain dripping from the palms and the tiled roofs. Les heads for the rocks beyond the river-mouth. I walk inland, along the river. I walk maybe a hundred yards, watching the brackish swell of the estuary. The sun warms me a little, and I stop. Squatting in the ankle-deep mud, I watch the stands of date palms, the brown wedge of the river, the ragged clouds headed for Guaymas, and the old fishermen pushing their boats from place to place. I watch the birds.

Nearby, Heermann's gulls and frigate birds are screaming at one another as they fight for small fish in the muddy shallows. Upriver a hundred yards or so inland a great egret stands stock-still in knee-deep water while a whimbrel pulls bugs from the mud. And on the near shore a snowy egret surgically skewers crayfish and soft-shelled crabs with its blackened bill. I, a dry-lander, am startled by this wealth, these birds. But it is January, and this dry little strip of land between two great seas is a warm wet place full of crayfish and beetles, full of fresh water and lush green palms, where birds wait while the ice gods have their sport to the north.

I watch the whimbrel for a while as it moves back and forth between the small risings and fallings of the river, feeding in the intertidal mud. Then I notice two birds in a mangrove across the river. The birds are large, nothing I've seen before. For only a moment they both sit in the same tree, then the larger one with a great squonk leaps into the air and charges the other bird and drives it across the river. The smaller of

The Rio Mulege, near Mulege, Baja California

the two birds settles on long black legs in the water five yards from where I am. She (I suppose) has a great yellow-and-black plume trailing back from her masked head. She is three, maybe four feet tall with a wingspan of five or six feet. For a moment we stare at each other, then, uneasy with our nearness, she lifts and settles a safer thirty yards upstream.

Three times while I watch, this bird moves back to the tree where the larger male (I suppose) bird is perched. Each time she is aggressively driven off with a great noise and show of wings and beak. Each time the female settles farther upstream. Finally, after a fourth rejection, this miraculous bird circles high above the river and flies inland beyond my sight. The male then—as though some question has been answered once and for all—flies off in the opposite direction. In a week or so I will learn that the birds are yellow-crowned night herons. But I am baffled by what has just taken place, the hostility of the birds in a place so rich in trees and water and fish, their relentless intolerance, their consuming anger, in this bay of little milagros.

Suddenly Les is standing next to me with a large smile splattered across his face and an even larger dog snapper sticking mostly out of his day pack. Les and his pack are covered with scales and fish guts, but he is beaming. He insists now that I join him. Impressed with the size of his fish, I agree. We return to our cabin, grab another one of his spinning rods (I only brought a casting outfit), and work our way back through the mud to the rocks near the sea. The tide is now moving in. Here the water rolls up yellow-green, all of the redness of the Colorado Plateau gone, yellow-green between black rocks thick with mussels and barnacles.

On the first cast I hook a small sculpin, a fish like a Japanese kite, all spikes and bright colors. Tasty, too, but potentially painful. Sculpin spines carry a nasty toxin. Les moves over and helps me pull the hook from the sculpin's mouth. I toss the fish back to the sea. And as Les complains, the sculpin is gone in a single orange spasm. I remind Les of the fish's size (small, even by his standards). He mutters something unintelligible and walks off down the beach, working the deep pools among the rocks near shore.

On my second cast I snag my hook on something solid beneath the waves. And then, while trying to jerk my line and tackle free of whatever I've hooked, I snap Les's rod cleanly in two. Les has been watching ever since it briefly looked as though I'd hooked a fish.

"Asshole," he shouts from down the beach, then again, "Asshoooole." The word hangs nicely in the cool sea air, then Les simply turns back to his fishing. Though a cheat, as I've told you, Les is a good man little bothered, in fact, by my gaffe. But neither is he likely to pass on an opportunity such as this.

"Asshole," he mutters loudly one last time, then moves beyond hearing.

I lay the broken rod where I can find it later and move off to beachcomb while Les passes beyond the world of mortal men and slips into the deep trance of the true saltwater fisherman.

I look for the first time at the beach behind where we were fishing. There is obviously a local fishery. Dead fish are everywhere. The people from the cardboard and plywood houses that line the lower river must seine the seas nearby, then come to this beach, no more than a hundred yards from their homes, to empty their nets and clean their catch. The sand is littered with the remains of their fish: many kinds of fish but mostly a small skate or ray, the sort you see for sale in Mexican tourist shops. A fish that when cut in a certain way and then dried looks like an especially evil representation of Satan—devilfish, or something like that, they are called in the shops.

Suddenly these rays are everywhere. I must walk upon them to move in any direction along the beach. Their stench is around me like a glove, warm and wet. The number of dead fish staggers me. Mulege is not a large town, a few hundred at most. But here is carnage that would honor many times that number, carnage as only humans can inflict it. I am reminded of pictures of the rice fields in Cambodia, or the hospitals and orphanages in Rwanda thick with human dead. I can't move without stepping on fish.

The people who slaughtered these fish could not possibly have eaten the flesh they took. Though some part of me knows that at least some of the meat flensed from these fish is eaten elsewhere, most of me just reels from the sheer numbers of dead, no matter the killers' conscientiousness.

For a moment the breeze dies completely, and I gag, hard. My mouth fills up with ropy saliva, and then I gag again. I place my hand over my mouth and nose, and as quickly as I can I move to higher ground, toward the breeze and away from the grim smiles splitting the rotting fish heads. Finally, the fish and their lidless eyes thin, then fall behind. Once I'm fully beyond the smell of them I stop and sit on a crumbling block of concrete with a stump of rusted iron pipe sticking from it.

Jesus, why do they do this? Why do they leave the fish heads to rot no more than a hundred yards from their houses? Leave them to rot where their smell and their blistering skins foul everything—the sea, the palms, the homes, even the dead themselves? Another hundred yards and most of it would be beyond sight, most of the time beyond smell. Is it just American prudishness? Or is there something more beneath the sun-ripened skin of these dark men with eyes like bruises, these women beautiful as flowers—something on better terms with the rich smell and the fierce eyes of the dead?

And what is it, in spite of all of this, or about of all of this, that draws me—like a moth to a candle flame, like a priest to little boys, like Van Gogh to sunflowers? And what is it that stirs in me even now, at once brutal and sensual? The sea just keeps folding and refolding itself, like brackish bread dough. The fish just stare at the sky.

Again, it begins to rain. Les, in his thin cotton jacket, continues to fish as though nothing were actually falling from the thick clouds, as though he wasn't soaked to the skin. I huddle beneath my rain jacket, listening to the BBs of water smack against the nylon fabric. And then for no reason at all, a marvelous thing happens.

Just beyond the beach, enormous cardon cactuses rise to heights of twenty or thirty feet. And as the rain increases, great black buzzards drop from the dark sky and settle atop these cactuses. There the birds spread their enormous wings into the wind and rain and just sit there, dozens of them—black like night, five or six or seven feet across, the wind lifting their primaries like fingers, the rain fallen from the dark sky washing their bodies, their wings, the red flames of their heads.

Les has seen them too. Together we stand and watch until we are drenched by the cold rain. The vultures, all facing into the wind, cling to their cactuses like dark blossoms filling with rainwater, like oily black orchids shuddering in the wet wind. For a long while we stare at the spectacle. But the rain is at last too much. Cold and drenched, we move off. The birds watch only the wind and the sea.

It is dusk now. The rain is still falling. I cannot shake the image of the vultures atop the cactuses. I wonder if they are still out there, bathing in the near dark. I wonder if any of it ever bothers them—the dead, their need, the rain.

Les and I, in dry clothes, are again playing gin rummy between where the leaks in the roof spatter against the tiles in the floor of our room. And again Les is cheating, I'm certain of it.

A Winter's Solstice

41.

It is the time of hunger,
the night of the lamp-eyed owl.
It is the place of sorrow,
the longest dark of all.

42.

Much of his life he'd spent
gathering bones.
    Bones of
    people mostly, but bones

of animals as well. He gathered them
by the barrow-full, and in his lantern-lit room, he

    sorted them, sorted them by size
    and by weight and by color.

The long marrow bones of the leg
and the arm in one box,

    the carpals and finger bones
    in another. The ribs

and the clavicles, he never kept.
Or the skulls, the sutured plates of the head

    pinned atop the curious curve
    of the buckled spine, these

he always left behind. He prized most
the tiny cruciform bones within the eye,

    the crystalline ones that focus
    the light, and the bones

of the ear— the anvil, the stirrup, these
he liked very much, and he kept them

in a coral box with a fish carved in the lid,
beyond the reach of the stinging black ants.

43.

I
am too long before the guardian,
too long here with my tongue
in the damp sleeve of its mouth.

44.

I shared a room once with a man named Paul.
A pipefitter with bits of burnt rod and grease
behind his ears, beneath his nails. Paul,
whenever he wished, could pull steel balls
from nothing but still air, or geese from bread pudding,
the sort made from cornmeal
and strawberry jam. I never saw how, but Paul
did these things with little effort and less thought,
and he did them whenever it suited him, he did them
as though fabricating geese were as simple as cracking
knuckles. The geese rarely stayed, but the balls
gathered in rusted heaps behind sofas and chairs, beneath
beds and desks and plastic dinettes. Because of Paul
and his balls, I finally left. I've neither seen nor heard
from Paul since. Neither, to my knowledge,
has Paul seen or heard from me.

45.

At the marbled mouth of Little Wild Horse Canyon, there's a cottonwood tree. A Fremont Cottonwood with bark, I think, like bones, bark like casket lumber. The tree is mostly dead, ruined by the routine drought here. When I was young and Catholic, Sister Agnes Eugenie told us that St. Peter had so grieved over his denial of Christ that his cheeks were rutted from the tears he wept. The ruts in this cottonwood are that deep, that perfect—the ruts in the rough and sexual skin of this old tree.

46.

It is the time of hunger.
It is the longest dark of all.

I've known rivers:
I've known rivers ancient as the world and older than the
flow of human blood in human veins.

My soul has grown deep like the rivers.
— Langston Hughes

Snowy Egrets: The Ending

A white Pontiac like a crippled steer staggers slowly from pothole to pothole down the rain-wrecked main street of Luis B. Sanchez. Beyond the Pontiac an old man is playing a concertina, another is singing. Both men are crying. And beyond them others, men and women and children, march solemnly behind the rusting automobile or simply stand in the two-day-old rain and watch. The silvered windows of the station wagon show nothing, but everyone here knows what is inside the car. A lucky man, the Pontiac's owner—space enough for making private love, darkness enough for a final dignity on the road to the camposanto.

The station wagon jounces slowly past my truck, then the mourners, then finally a string of cars, some with headlights on, most with no headlights at all. And then abruptly all of it vanishes down a side street. An instant of intense silence, then the rain begins again. And suddenly the sounds and smells of the village of Luis B. Sanchez—frenetic Mexican Spanish, tortillas in hot oil, frijoles and salt pork, roasting peppers, dogs barking, and as always hereabouts, the rich smell of raw sewage.

For a moment longer I sit with the rain sleeting in my window and think of the corpse. For no particular reason I imagine an old man, brown and deeply creased from days beneath the fierce sun in the kale fields north of here. And then I think of the gift of his friends—men and women who've slept with him, eaten and drunk with him, prayed with him—gathered here to mark his death. Friends are mysterious and powerful things, even at the very last.

I tip my cap in appreciation, then raise the glass against the rain and drive on. I'm looking for Baja California Highway 1 out of Luis B. Sanchez, the road west to Ledon and Jesus Gonzales Ortega. It seems the Mexican people either don't travel much or their skill for dead reckoning is uniformly exceptional. Except for detours, and then only where they begin, I haven't seen a road sign since San Luis Rio Colorado, south of Yuma.

The dirt and stone strip I'm following ends in a T. I turn to the left—six of one, as they say. The road becomes a deep bed of rain-soaked sand, lined by those lethal eighteen-inch-high curbs you see only in Mexico. Though visions of quicksand fill my mind, the sand, I rationalize, is perhaps a good sign. Two long blocks later I cross another road running east-west—a paved road!

In nearly all of northern Mexico the pavement alone would be enough for positive identification of the highway. But unlikely as it seems, two paved roads leave little Luis B. Sanchez—B.C. 4 and B.C. 1. Either would work, I guess, but I'd prefer B.C. 1. I stop and look at the map. I've no idea where I am. But, according to the map, B.C. 1 heads due west from Louis B. Sanchez. B.C. 4 heads northwest.

A woman is standing at the corner, waiting for a bus. She has been discreetly eyeing me since I stopped, wondering I guess just what the hell I'm doing. In my broken Spanish I try to ask if this road is Camino Uno. To my amazement, she imagines I speak the language, and she responds to my question with a complex set of instructions, none of which I understand. But she finishes by pointing toward the west.

I thank her for her kindness, though I've no idea what she said. For all I know she may have told me where I and the horse I rode in on could stick the whole history of U.S.-Mexican relations. I turn west anyway—I mean, do I really have a choice?

For the first few miles I watch the compass carefully. A little south, and then the needle, led by the honest-to-god magic of geomagnetic flux, swings to the *W* and is pinned there. I have no idea where the other road could have been. I feel certain I saw nearly all of the town back there. But according to the compass I can only be following Baja California Highway 1. I relax. Nothing left now but ten miles of rotten asphalt and murderous shoulders between me and the last brackish reach of the Colorado River.

The land here runs so evenly to the horizon that the bottoms of the leaden clouds are hidden beyond the shallow curve of the planet. The topo maps of this region are bare as a baby's butt, unmarred by contour lines. Even a west Texas flapjack at two hundred yards would block a sizable part of the view here. Euclid himself would have liked this part of Mexico.

The altimeter reads dead zero, and all around it's green. It is February, but I have come to where the goddess of summer waits while ice rules the north.

Four days ago I sat in eight or so feet of crusted snow atop Loveland Pass—two miles, plus 1,402 feet above where I am now. From up there I surveyed the first verdant reach of Colorado River drainage, the valley of the Snake River. And, probably because of oxygen deprivation, I determined I should see the rest—all 1,450 miles—of that drainage. So I pushed my truck into neutral (1,450 miles is a long way), let off the brake, and rolled on down the western face of the Divide.

From Loveland Pass to Dillon the road hugs the North Fork of the Snake River. As in many places, here the river did the early hard work—the road just follows its lead. At Dillon the Snake joins the Blue River, in from Breckenridge, and together they spin off to the north toward Kremmling. Along the way the Blue River gathers up all the water between the Gore and Front Ranges of Colorado then dumps into the dark gorge of Gore Canyon.

Just beyond Dillon, though, I climb up onto I-70, leave the river behind, and head west toward Vail Pass. Interstate 70 follows the fragile little flow of Tenmile Creek draining this side of Vail Pass—still Colorado River water, but paltry and oddly flowing east. Once over the pass the water catches the first true westward tilt from the Continental Divide. And on the left, then on the right, then on the left again runs Black Gore Creek, then Gore Creek. Then just beyond Vail the Eagle River rolls in—down from Tennessee Pass near Leadville, where the drainages of the Colorado and Arkansas Rivers divide.

The Eagle's the first genuine river along this stretch of road. Not like the Colorado mind you, but broad and angry. I've seen golden and bald eagles here, too, and elk and weasel and deer, and it's rumored that the moose are moving back. A good river.

Still, I am, as always, warmed when I get well beyond the wound of Vail and into the sagebrush flats near Gypsum. I like Gypsum. Not because of all the calcium sulfate they bottle here, or the mess it makes, though I am partial to messes, but because just beyond Gypsum is Dotsero. And though I can't say much for Dotsero itself, this is the spot where the Colorado River finally crashes out of the Rockies, shakes off all of those pine needles, and lurches off to take another swipe at Glenwood and all those other downstream canyons. Plus, Dotsero is where I'm reunited with all that water I left behind when I parted company with the Blue River back in Dillon. So I'm always pleased to reach Gypsum.

It's rivers I guess—though combined they contain less than one one-thousandth of 1 percent of the water in this world—*some thing* about that water and those rivers nags at me. And it is especially the Colorado—a river as fair and broad as the lustrous buttocks of Botticelli's Venus, a river that gathers all of the rain and snow and sweat and urine that falls as far north as Mammoth Glacier in the Wind River Mountains and as far south as Ojo de Agua in Sonora, everything liquid from Montana to Mexico, from the Rockies to the Sierra Nevada—the Colorado River that is, that keeps dragging me down this road.

At Dotsero I stop to taste the waters of the Eagle and Colorado Rivers. Icy here—tastes of things underground and bawling winds over vertiginously corniced ledges of rotten ice, cold rocks, and goat shit. I urinate where the rivers meet. My contribution. Then I walk back along the gravel shoulder to my truck.

From Dotsero to Grand Junction, out through Glenwood Canyon and the Piceance Basin, the road snakes alongside the Colorado River. I roll with the water, deep green through Glenwood Canyon, mottled by ice this February. And as I always do here in the shade of the splintered limestone, I think of friends. Friends I will likely never see again: John and Martha and those long trips in VW microbuses to the canyons of southern Utah, tents full of beer and burning hemp; Marjorie and her rose-tipped breasts and blood-stained sheets. And friends who, because of what I've become, will never be friends again—not in the same way anyhow—friends I drank and got laid with, friends I learned from, friends I grew up with. And as always, something beneath my stomach twists on itself. I resent the indifferent warp of the years. It seems an odd

Friends

The "Gunderson" River, near Grand Junction, Colorado

expenditure of energy in a deeply entropic universe for friendships to burn so brightly but so briefly. I miss those people, those times.

Solitude is a fine thing. Fine, that is, so long as it's voluntary.

Outside Grand Junction I leave the river and head south toward Dove Creek and the Devil's Highway. Just beyond the city the road crosses over the Gunnison River, or the "Gunderson," as my brother Michael calls it for a reason I can never quite recall. I stop again and walk down to the river. The Gunderson comes from the western slope of the Rockies and the high plateaus of Blue Mesa. Here it tastes of glacier grit and very faintly of the huge mackinaw working the dark rifts at the bottom of Blue Mesa Reservoir. The river helps push the faces of friends from my mind, helps unite that ring-bend in my stomach, and I drive on.

South of Grand Junction the road runs through a crack in the Uncompaghre uplift and the easternmost edge of the Paradox Fold through the heart of the Colorado Plateau. A fine drive, even in winter. The road follows East Creek through Unaweep Canyon to the Unaweep Divide, where the drainages of the Gunderson and the Dolores separate, and then follows West Creek down to Gateway, where it joins the Dolores River itself.

Unaweep Canyon is a fossil of sorts, a fossilized footprint most likely left behind by the Colorado River. This canyon was once the bed of a great river, probably the Colorado, maybe the Colorado and the Gunderson, before a cataclysmic uplift forced the rivers north. Even to a dilettante geologist like me it's obvious that the meager flows of East Creek and West Creek didn't carve this glorious crack in the Uncompaghre uplift. So here's a little trip down memory lane, a little glimpse of where the Colorado walked when men and women were ferns.

After Gateway the road winds along the caramel-colored defile of the Dolores River canyon. A little out of the way perhaps, but there's a fierce solitude here, and this road allows me to skirt the developmental tumor swelling around Moab.

Why is it, I wonder, that I rarely make close friends any more? Is it because I'm less available to people than I used to be? Or is it because I have less time or less appetite for people? Or is there something important about being alone as we age? I hope not. Do other people waste their precious time like this? Maybe it's just that I'm a cantankerous old prick that people don't much like being around. That's probably it.

A magnificent red-tailed hawk swirls low over the river, and his honest-to-god indifference pulls me back to the moment.

Frie

The Dolores River, near Naturita, Colorado

Where the road next climbs down to the river I stop and get out the tuna salad sandwich and Fig Newtons I bought at a gas station in Grand Junction. I sit by the water and eat. The shredded muscle inside the sandwich brings to mind the silver spasms of yellowfin tuna beneath the Sea of Cortez. I thank the fish and apologize for the exigencies of life. But, my finned friend, your turn will come.

Then I eat the fig bars and curse R. J. R. Nabisco. I know it's wrong, but I can't help myself when it comes to Fig Newtons. I throw rocks and sticks into the river and watch the wood work its way slowly north. North?

The Dolores River here completely violates my sense of direction, sense of tilt, and seems to flow uphill. From Dolores, Colorado, the river turns north and runs near one hundred miles that direction until it joins the Colorado near Dewey, Utah, just north of Moab. From there, the Colorado River turns due south and retraces that hundred miles no more than fifty miles west of here. That seems somehow wrong, seems wasteful. But water only knows which way is down, which way is easiest, my sense of propriety be damned.

The Dolores tastes a little of the snowfields of the San Juans, a little of glacier and avalanche lily, a little of lupine and larkspur, a little of, yes, spearmint. That's good. But mostly, the Dolores tastes like mud. Good red Colorado mud, but mud just the same.

At Uravan, the road leaves the Dolores and joins the San Miguel River, recently in from Telluride. Then beyond Naturita, near Nucla, where great sportsmen and ardent environmentalists used to gather each year to make fools of themselves, the road heads south and curls down into the Disappointment Valley. A disappointment for some perhaps, not for me. To the west rise the fair faces of the Manti La Sal, snow-covered now. To the east the sacred forests and peaks of the San Juan Range. And as though this weren't enough, to the southwest the Abajos—the Blue Mountains—rub their shaggy backs against the sky.

The Disappointment Valley, near nowhere, Colorado

Disappointment Valley would be a fine place to die, a fine view to hold for eternity. I think for a while of my wife and me lying out here while the pink flesh fell from our bones and wood rats chewed our toes, while our hair grew long as snakes. Yes, a good place to lie for a long, long time. But not today. Today only the spirit of my wife is with me. And today I have a river in my head.

Then, all at once, the Dolores reappears briefly, my truck noses up, and I begin the 2,100-foot climb up Slickrock Hill onto the Four Corners Platform—a chunk of rock pushed up in the Triassic. A sudden reminder of the awesome geomorphology of this continent and the irresistible fist of gravity—gravity that turns all of those turbines, pushes all of those rafts, gravity that holds the moon and lifts the seas, and gravity that ripped the guts from the Kaibab Plateau and left them strewn all over Sonoran Mexico and Baja California. Powerful shit, that gravity. I make myself a note to be more grateful to Newton, and I try again to forgive Señor Isaac for the curse of calculus. Again I fail.

Great view from the top. Then a short, straight run, towns with names spelled backward, then Dove Creek and Highway 666 into Monticello. A quick left there and all too soon I'm in Blanding, Utah—a town at least as friendly as its name. I stop for dinner, a bowl of chili made from knuckles, I think. Yep, knuckles from some old steer or unfortunate schnauzer. And then some of that famous southern Utah hospitality where a bill and a map of the route to the door come with every meal. Then, at last, a warm bed.

I am well away from the river here, but as I lie between the hypochlorite-tainted sheets I can hear, yes, there it is—where it grinds at the rocks and the dams, where it wets the tongues of lion and deer and tamarisk, I can hear it, and it lulls me.

Up at the crack of nine, I'm off again for the river. First, though, I am bound for the improbable stone sombrero of Mexican Hat and the boiling brown back of the San Juan River. I grab a couple of prefab egg salad sandwiches and another box of Fig Newtons at the Minit-Mart, then head south, sorry this day that I must pass all of the roads heading west.

Near here, two hundred Mormons blasted and chiseled their way through Hole-In-The-Rock and cabled their oxen, their women, their carts, and their children across what most believed was the impassible divide of Glen Canyon. And they did all of that just because Brigham Young asked them to. And when they were done, their reward was the town of Bluff, Utah. Mighty susceptible to suggestion, those Mormon pioneers, and terribly persistent.

And to the east, the poverty of the Hopi and the horror of the Hanta virus. Deer mice may yet even up a few of those old mammalian scores. And just a little north and east of here, near Hovenweep, my first wife and I spent a night in an old sixteen-person canvas tent with wooden poles thick as Hulk Hogan's wrists, smoked and drank and talked about the ferocity of the stars and the fierce red pain of solitude. Years later that pain would rise from the exhaust of an old Subaru and grant her a final solitude. The river is close now, I can smell it.

The San Juan River at Mexican Hat, Utah

I work my way down to the San Juan near an old suspension bridge just out of Mexican Hat. East of this bridge on the south side of the river, some strange-looking conglomeration of machinery—perhaps because the freezing weather has burst a pipe, or perhaps because whatever this machinery does, this is simply the best way to do it—is spraying water from some twenty or thirty feet above the river onto the rocks below. Where the water hits the rocks it has frozen into shapes like toadstools and stalagmites, shapes like the hands and the feet and the fingers of Venusians or Jovians. What is it about Jupiter, anyway, that makes its citizens so jovial? The water falling from above isn't colored, as you'd expect sewage-tainted water to be, but I move upstream for my drink just the same.

The San Juan River runs an odd course south from Pagosa Springs and north from places like Chaco Canyon and Pueblo Bonita and Shiprock, then north again through the broken toes of the southern San Juans, and then finally west, snaking and goosenecking its way among the slickrock canyons of southernmost Utah into Lake Powell. It's a popular float trip.

I bend, fill my hand with its brown-green water, and pour a little over my lips. A damn fine river this San Juan. Tastes a little magical today. Must be the silt from New Mexico or Arizona. And faintly, very, very faintly, but unquestionably, I taste grizzly bear piss.

Indian "rock art," near Mexican Hat, Utah

A little more than twenty years ago a grizzly sow was shot and killed by an outfitter out of Durango. Some think it likely that sow had cubs. Apparently those cubs, their aunts, or their uncles, or, I hope, all of the above, are still out there.

It's a glorious day, nothing but sun in the sky and nothing but the rumble of the ice where it nudges its slushy flesh against the stone river channel. A damn fine day. On to Arizona—the Grand Canyon State, you know.

The buckled and broken spires of Monument Valley rise like some series of lewd gestures from the mittened fingers of the desert. Perhaps it remembers after all. But, even the stones' defiance, even the whorls of desert varnish and the light shattered into a million shards by the crumbling buttes and mesas, even all of this is not enough to dispel that odd feeling that gathers itself somewhere behind my belt buckle as I cross into the Navajo and Hopi Reservations of northern Arizona.

For me at least, it is a strange thing to see such struggle in the midst of a desert so wholesome, so nearly perfect, a desert where the likes of John Wayne, The Duke himself, brought women and children and cattle through all kinds of trouble, a desert that at times fills up with moonlight like a sacred stone cup. Somehow it's different in Mexico. Maybe I'm just used to the dry unshaven face of life there, or maybe I feel less responsible for what's happened to the people of Mexico—though I'm not. Maybe I'm just afraid of being a bleeding heart who knows all the right words but doesn't actually give enough of a shit to do anything about it.

Then, Kayenta, Arizona—with its manila-colored military-style homes, its skinny dogs and rusted pickup trucks, its serious sunburned people—smacks against my windshield like an old paper bag, cutting off my view. And the seams around my windows fill with red dirt. I try to drive on, but it's about 3:30, and again and again I am stopped by the flashing signals of long yellow Bluebird school buses. As each braided and beaded boy or girl careens off the buses and spirals off down one of the red dirt roads into the juniper thickets, I wonder: Where are the homes? What are evening meals like? Does the warm palm of this land make up for the cold bone of poverty?

I pass under the grim green coal chute of the Peabody Company. And I think of the polluted or dry wells, the endemic asthma, the barbed-wire fences, land that will now grow only sparse hard grass, and the dollar bills flowing into the pockets of people who've never seen Navajo Mountain, a desert sky, or the fire of globe mallow in the spring.

Finally the buses disappear and the land begins to roll past beneath my tires. Tsegi, Cow Springs, Tuba City, then U.S. 89 and a blessed left turn. Near where the Little Colorado crosses beneath the highway just outside of Cameron I pull to the side of the road near a display of silver and turquoise. "Welcome, Best Deals!" the plywood sign says. I slow, then stop. I open my door as though to let someone out, take a long deep breath of air syrupy with piñon and juniper, breathe out, breathe in once more, and drive on.

Then I begin the slow climb onto the Kaibab uplift, the climb to Grand View Point and the fearsome pink and brown and blue and black spires, the knobs and chasms, ripples and shelves, fissures and buttresses, nipples, vistas, and just plain god-awfulness of the Grand Canyon of the Colorado River.

Snow is beginning to fall as I pull into the Village, and where I leave the road the earth seems cleaner, seems wider, seems almost serene. For as long as there is light I stare into the canyon, across its terrifying breadth and up the girlish gash of Bright Angel Gorge. Above, the visceral green of the piñon forest stands in stark contrast to the surgical white of the snow.

And then as the light finally fades, in the darkening face of the Redwall, I imagine for a moment that I see images. Images, perhaps, of people momentarily familiar, then the limestone sinks into indistinct darkness. The snow is falling heavily now. It is good to be near the river again. Ravens still push at the shadowy air as I head off to my cabin and bed.

And it is the ravens that wake me, calling to each other—Hey! Harold! Hey! Carl! Hey, Callahan! Get your lazy ass out of bed! At least that's how it sounds this morning. Though the sun has risen, it is cold and the canyon is filled with a thick fog—cream of river soup—that boils up in ragged wisps of hoary lace. The gauzy breath of the river itself, come for me.

I dress, then submit once more to my longstanding lust for luxury and breakfast at El Tovar Lodge. Basted eggs over artichoke hearts and puff pastry filled with spinach and hollandaise sauce, thick rich coffee, and orange juice full of pulp and seeds. I eat slowly beneath the black-beamed Oregon oak ceiling and splotch the neatly pressed white tablecloth with sauce and spinach and coffee. I watch the canyon beyond the window. The murk opens and closes likes a woman's skirts, tantalizing glimpses of red flesh and the deep folds of this earth.

The waiter has been kind, but by my seventh or eighth cup of coffee even my stunted sense of guilt is roused. I leave a big tip . . . well, for me it was a big tip. And I head off, with caffeine roaring in my veins, toward Williams and Flagstaff, toward Yuma and Louis B Sanchez. On my way I toss a stone into the canyon, and though I can't see because of the fog, I'm pretty sure the rock hit the river below, because of the splash I heard.

One of my favorite places in the whole world is Quartzite, Arizona, especially in January and February. Quartzite sits at the intersection of Interstate 10 and U.S. Highway 95. For two months a year, Quartzite becomes a scene from *Road Warrior,* or *Prospero's Books,* maybe. Each year in January and February, especially on weekends, exactly 8 million people, each with his or her own motor home, converge on Quartzite for the annual gem show, international lunacy competition, and world's largest swap meet/flea market/auto upholstery/furniture/pottery/gun/boat/and fireworks show (not to mention the sideshows with variously altered humans and dogs). Mostly the 8 million people here just drive their Winnebagos back and forth on Quartzite's one street, possibly looking for a place to park, but mostly I think, just for the spectacle of it. A few others simply put on strange clothing and walk up and down the same road, exchanging pleasantries, business cards, what have you. As I said, *Road Warrior.*

It takes me fifty-five minutes to transfer from Interstate 10 to Highway 95, a distance of some six or eight blocks. But the show of humanity—way more than two standard deviations out from the norm—repays me handsomely for the delay. And once I head south again, the traffic thins quickly.

The drive from Quartzite to Yuma is one I truly enjoy. The river is beyond sight to the west, behind the Chocolate Mountains, carving out the boundary between Arizona and California. In fact, the changing course of the Colorado here continually redefines what is Arizona and what is California. Squatters have for years taken advantage of the consequent questions of jurisdiction and standing. I admire that.

And it used to be that south of here the river almost annually changed its mind about where to empty and how to get there. Some years the river flowed west and north through the Alamo River to revitalize the Salton Sea in California, other years through the New River, and some years through the Bee River. Eventually, each of these channels would silt up with Utah or Arizona soil and the river would again reach to the Sea of Cortez, for a while. That was before human hands took charge of the river's course. But even as recently as the spring of 1905 the river rose up, ripped through a dike in Mexico, and turned due west. For the following two years the Colorado contentedly emptied its entire contents back into the California sink, adding considerably to the girth of the Salton Sea and the tourist appeal of Borrego Springs.

In the winter of 1907, aided by unusually low flows, Southern Pacific Railroad crews were able to haul and dump enough rock, brush, wire, dead cattle, Buicks, and what have you to force the Colorado back into its original channel to the sea and restore Colorado River water to the Imperial Valley (once known, for still obvious reasons, as the Valley of Death). Twenty or so years later Boulder Dam was built, and the silt and the semiannual flooding that had added such intrigue to the yearly course of the Colorado were eliminated. Too bad. But rivers are persistent, persistent and more than a little crazy.

East of Highway 95 rise the volcanic pinnacles of the Kofa Mountains and the unbroken sagebrush flats of the Kofa National Wildlife Refuge. Every time I pass through here I swear I'll return with time to explore these magical black turrets and minarets, these vast stretches of humanless land. But today, again, I am full of other purposes.

A little north of Yuma the desert suddenly goes from pale brown to deep green, palm trees spring up, and children and dogs appear at the side of the road. And even if you know little of this country, you begin to suspect something wonderful is about to happen. And, in rainy years, anyway, it still does. Out of the godless waste of southern Arizona, like blood from stones, there is suddenly the wet gift of the Gila River. With water gathered from as far away as New Mexico the river rolls beneath the highway, anxious for its marriage to the Colorado.

I stop to sip from the last United States tributary of the Colorado River. Tastes of the white rocks of Aravaipa Canyon, the rusted rhyolites of the Chiricahua Mountains, the toilets of Phoenix and Tucson, lizards, snakes, sheep shit, and scorpions. I watch the cars passing. No one else seems even to notice the Gila River. I wonder, as I watch the Fords and Chevys, the RVs, and the pickups roll across the bridge without slowing, if I am the only one who sees this river, the only one who suffers from this water thing. Tomorrow will see the last of it.

There is a bad smell in Yuma, Arizona. Not one you notice with your nose. The air is fine. There's something else, though—something you smell with your skin, something I've smelled nights in jails and whorehouses, something bad, something that makes me wish things were different than they are.

I fill my truck and my gas cans with unleaded, not sure how far the end of the river is from a Pemex station. I buy a couple of maps, which I'm sure will prove useless, and some car insurance, which I hope will prove unnecessary. Then, reluctantly, I get a room for the night. A room in Yuma. I drink cheap wine until I can't smell the smell anymore. Then I sleep.

Early next morning I cross over the Colorado on Interstate 8. Beneath the roadway the river is confined to a narrow concrete chute, and all of the water allowed by international law is siphoned off here into the All American Canal—water for the Imperial and Coachella Valleys, North America's salad bowl. The United States leaves just 1.5 million acre-feet of the Colorado for Mexico—exactly what was agreed to in the Colorado River Compact of 1922—no more, no less. Hey, in the American Southwest, water is gold—U-period, S-period gold.

I leave I-8, and take the last of U.S. 95 south toward Rio San Luis, Colorado, and the great nation of Mexico. It has begun to rain, not hard or cold like in Colorado, but a steady rain that glazes everything and turns the road's shoulders to swirling mud.

By this point, most of the Colorado has been used to irrigate one piece of land or another, has passed through one pair of kidneys or another, and has briefly been blood, sweat, tears, or semen—or all of the above—somewhere upstream. And in particular, the water has been used to irrigate the grapes and the melons of the Grand Valley, a valley lined like a tub with subsurface salt zones. And what's left of the Gila has been used to irrigate the Wellton-Mohawk Project—a BuRec idea for agricultural reclamation that generates waste water so salty that most of it has to be pumped directly through Mexico to the Sea of Cortez. Add to this evaporative rates that would impress even the Saudis, and what's left of the river is salty, some ten to twenty times more salty than the headwaters. Currently, though, the water isn't salty enough (more than 115 milligrams per liter) to violate Richard Nixon's 1973 agreement with Mexico. But that's only because rainfall and snowpacks have been high, plus the state of Arizona has not been taking its full allotment of river water.

If snowfall in the Rockies or the Wind River Mountains drops, or if Arizona gets a little thirstier, or all three—then the water from the Wellton-Mohawk canal would have to be added back to the Colorado to produce the acre-feet of water we've agreed to ship to Mexico. That, of course, would send the salt levels back to where they were when Mexico threatened to take the United States to the World Court. So the only way Wellton-Mohawk water can be used now is if we desalt it. To meet that eventuality, the Bureau of Reclamation has built a desalination plant in Yuma. It sits now, fully operational but silent—waiting with its machinery, men, and women for the day the river turns toxic. And it will—given birth rates and climatic changes, it definitely will.

In San Luis, just before the border, I stop for some breakfast at a McDonalds. Okay, I know it's wrong, but sometimes it's just simpler (if you can bear the guilt, and I'm getting better at it).

Border towns fascinate me. U.S.-Mexico border towns are mostly filled up with people recently Mexican—they can muster a hell of a lot more enthusiasm over minimum wage and regular hours. Spoken words are usually Spanish or a lyrical mixture of Spanish and English. The sheer foreignness of it—the smells, the sounds, the words—and the raw vitality of it all make my skin tingle. It also scares me some, because I'm mostly uncomfortable being around large numbers of people, even people whose words I understand. Here I understand almost nothing. But you'd think there would be one advantage to all of this. Because even the U.S. side of border towns is so unlike most other American towns, you'd think you'd hardly even notice when you cross the border, but you do. Least I do. Always, and abruptly.

The people who want to tint your windows, reupholster your car, or simply wash your windshield for a buck swarm along the sides and in the middles of the roads. Cars in every stage of ruin are everywhere on the streets, burning oil, grinding out old bearings, spinning fan belts—horns blare, children scream, and knots of men smoking cigarettes line the sidewalks, where there are sidewalks. Smells of frying tortillas, burning garbage, and broiling meat fill the air. The rules are different here, way different. You notice that immediately. The rules are different here because they have to be, and you notice that, too.

I dodge the entrepreneurs and look for road signs marking the road to Golfo de Santa Clara—not my goal today, but the shortest route to Louis B. Sanchez and the last reaches of the Colorado. I find the route, take a quick left around a stalled '65 Oldsmobile, and move quickly out of town. The air thins, the traffic dies out, and suddenly it is nearly quiet. My pulse slows and respiration begins again.

The river is beyond sight now. But to the west of here is the last dam on the Colorado River—the Morelos Dam, built to collect water for the farms and fields that surround Mexicali, Baja California's second-largest city and most important agricultural basin. Most, sometimes all, of what's left of the Colorado River is shipped to Mexicali via the Alamo Canal.

In a curious hydrological and philosophical twist, much of this water once again returns to the United States and the Salton Sea via the New River. I think it is paradoxical that this part of the river's water comes to rest where the Colorado and the Pacific met long ago. And I think it is interesting that along the way the water gathers tons of toxic effluents from the U.S.-run maquiladoras operating in Mexico to avoid United States restrictions on dumping. Now, when the wind is blowing across the Salton Sea from the east, you can smell our offal more than thirty miles away in would-be resort communities like Borrego Springs and Salton City. This stench has so far squelched plans for further resort development in the beautiful and delicate deserts east of San Diego. Water will have its way.

Below the Morelos Dam, in wet years anyway, the last of the river races for the Sea of Cortez. Here the land is flat and filled to overflowing with silt—sand and soil scratched from all of those upstream red-rock canyons, scrubbed from the stony bottoms of arroyos and cuestas, sliced from the brown backs of the bajadas and the badlands—full to the brim with sand and flat fucking empty of reasons.

For the next 750 or so years, the flow of that silt will be stifled, mostly by Glen Canyon Dam. But then, around A.D. 2700, levels of mud in Lake Powell will near the high-water mark, and silt will begin to move through the penstocks of Glen Canyon Dam. Boulder Dam will slow it again, as will the smaller dams below Lake Mead, but for fewer years than Powell. And within a millennium or so the rich red blood of the Colorado will again spill across this fertile delta at the river's mouth, and once again this fair land will blush with the rise of new life and a patina of old stone. Water is mostly filled up with patience. I am less so.

Twenty miles or so of a detour rotten with rain and I find the turnoff for Louis B. Sanchez. For a westerner, it is hard to even imagine a place so green.

That was this morning. It is good to be in Mexico today, good to be out of the border towns and away from those grim green Chevy Blazers and Dodge Power Wagons that the Border Parol seems so fond of, and it is good to be beyond most human habitation and once again near the river. It is thick here with new life, and it is still raining.

The road, Baja California Highway 1, bends back to the north, and from the map I know I'm close. I strain for a glimpse of it. Then, suddenly, the river. Sooner and different in nearly every way from what I imagined. If there ever was a bridge here, it went with the heavy rains a year ago, or the year before. Sand, gouged from the heart of the Colorado Plateau, has been scraped up from the river bottom, and a narrow raised causeway now reaches across what was once the river. As far as I can see there are tamarisk and low willows, the river's bed. God, it must have been magnificent, miles across. Miles!

I cross a little flowing water confined to aluminum culverts at the southern edge of the great channel. Then, following a blackened bus, I drive on—looking for the main body of water. The road rises and falls across the many channels cut by the river. Down into the soft sand, then up over little humps covered with tamarisk, down again, up again. But there is no water here, only garbage—an abandoned Dodge, rusted washers and refrigerators, black plastic bags full of rags and Pennzoil cans, useless toothbrushes and tires, broken metal chairs, a pair of men's underpants, televisions with their screens kicked out, a burned blue sofa, a doll's head, a broken lightbulb, a yellow hairbrush—only garbage. Then pavement again, and the first fields and garages outside of Ledon.

Jesus, it must have been that first water. I turn the truck around and drive slowly, almost reluctantly, back. When I cross the aluminum culverts I pull off the road into the sand and get out. The river is confined here by three thirty-six-inch-wide corrugated pipes cemented in place with sand from the great gorges of Colorado, Utah, and Arizona. In each of the pipes the water rises no more than six or eight inches—about as much water as there is along the uppermost stretches of the Colorado River Trail beneath the stolid gaze of the Never Summer

Mountains. Nearly dead here, twenty or thirty miles from the sea. Nearly spent for the lettuce and the sorghum in the Imperial Valley, gone for the lights of Phoenix and the lawns and fountains and toilets of L.A. and Orange and Santa Ana, nearly emptied for the palms and the bougainvilleas and the pickleweed of San Diego and San Bernadino, almost drained. Clearly, in dry years the river wouldn't even make it this far

By reflex, for I surely wouldn't have done it if I had thought about it, I stick my hand in the water and raise it to my mouth. Oddly, the river tastes fresh and cool. But I realize the risk, and spit the water back.

The bus from Ledon to Luis B. Sanchez and beyond crosses the Colorado River near its end

I expected salt from the sea and sulfates and nitrates and 2,4–D from the fields north of here. Maybe it's the rain, maybe it's low tide, maybe both, but the water isn't brackish or bitter.

I imagine the great tidal bores that once roared through here, walls of water gathered by the moon and sun, large enough to overturn seagoing ships on this, a once-navigable body of water, navigable as far north as Parker! Unimaginable now. And there were forests, equally unimaginable.

I step into the thin flow of the river, unzip my fly, and share my water for a final time with that of the Colorado. Four years and 14,000 feet ago I stood in the same posture up where the river is born. There is still within me a deep sense of alliance with this water, still a vision of the single thing this river has become, the single thing we have become.

But the images, like the waters I'm standing in, have been changed by all of this, changed in many ways.

I close my pants, and I turn to see where the river goes from here. South. Shells are everywhere along the shore—little bivalves mostly, validating my proximity to the sea. Then I notice something else. I notice that each of the dry branches and bits of board that line the river's banks look familiar. And I suddenly recall setting each of them afloat upstream. Now here they all are, gathered in the sand at the river's end, every one of them.

Beyond that, near where the last of the Colorado spreads dark with a reflection of the Sierra Major, there are egrets. Twenty or thirty, maybe a hundred, snowy egrets rising from the mirrored mountains on black stilts and scratching food up from the river's mud. Egrets white as the ice spires where the rivers divide, white as the fires that burn inside of stars, white as salvation itself—egrets like monks raking the river's bottom. For a while I watch.

Then I reach with my wet left hand and pick a crab's claw, red and tan, from the sand at my feet, drop the broken claw into my pocket, and walk back to the truck. I open the door and settle onto the seat, leaving the keys in my pocket, the door open. I listen to the egrets croak. It is raining again.

In Luis B. Sanchez, in the rain, an old man is lowered into the earth while his friends and his priest watch. Each of those gathered is praying for something different—resurrection, an end to pain, a child's health, forgiveness. Most of it can never be, of course. For that is the nature of things. But we have no way of knowing which it is we pray for—the essential or the unforgivable, for that is also the nature of things. And that is good.

Watermarks

When I was young, and in a place where no one should be,
feral water rose up and struck me. There are marks yet
where it touched me, and where it didn't.

I think the water remembers, knows me still by those marks.
Whenever I have looked closely, I've found that waters still bear
my image, still mirror my curious frown and uncertain thickness.

In brackish waters, where the warm liquid washes the palps
of stalk-eyed lobsters and cradles the shards of broken bone fish,
the image is most clear, most vivid. Because it struck me,

because I was very young, those who do not know water expected
I would die. Of course, they were right, but having for a long time
watched water, I no longer think that death is what the others say it is.

*

That winter, the cold fell upon us like a fist,
and the pond froze clean through, froze so clearly you
could see the fish fixed in the ice—the perch, the catfish.

You could see their flat black eyes set in dead stares,
their fins fixed in final waves, and the barbels on the catfish
looked as sharp and as dangerous as people always said they were.

I watched those fish all that winter, and wondered about the thoughts
frozen in their gray ganglia at the moment the ice gripped them.
Were the thoughts still there, trapped in the brittle blue ice?

That spring, as the pond warmed, I watched intently, waiting for some insight into that frozen instant, but as the water round each of them thawed, the fish simply rolled onto their backs and floated slowly to the surface.

*

The night we caught the moray, twisted in 20-lb line like Ahab's arm,
the sea reached out for us. A viperous tide uncoiled by the moon,
rose up through the rusted barnacles and broken mussels,

rose up round the pale flesh of our calves and pleaded
with us to walk with it, to ride upon its back to the sweet dark
heart of the sea. And for a moment, we wanted to go.

But the songs of our wives called to us, and we were yet of the rocks
and the sky. We walked from the water, the eel on an iron gaff stuck
through its rubbery head. We hid the fish in a cracked foam cooler,

among the bait and bits of scale. At home, in the cold of a night-white
porcelain sink, mouth full of needles, the eel struck at us,
like a broken snake still full of the fury and the grief of the sea.

✳

Here, full of the fierceness of June, the South Fork raged
through a granite defile and raked its blue teeth
cross the toes of the great rock peaks. At such times, in such

places, I am driven to throw stones and set pine boats to fair sail. . . .
Once, in Glenwood Canyon, I saw where they had diverted the great
Colorado from its channel. Where the river should have been,

there was only the bowel of the mountain. I'd always imagined other
things beneath rivers—gold and porcelain fish, tortoise shells, the brass
bones of dancing bears, and grass thick round the iron teeth of old gods.

Here, there was only the gray-green rock. . . . Once, near Jesus Gonzales
Ortega, I walked to the River to look for those sticks I'd thrown into
these mountain streams. They were there, of course. But then, I knew
they would be.

✳

The sacred springs of Guadalupe rise among the volcanic ruins of
the Sierra Juarez. They boil through the iron crust of the desert
and flesh bone-white agave and fierce green palms.

The Cocopah ate of these waters—black coyote, red deer,
white-toothed wood rat. The rocks here are scarred with their tales.
The hard edge of consciousness is most silken in water,

and the coarse flesh of friendship most evident in empty canyons.
Canyons where the false blue of the desert night
falls from the darkness like the lion upon the fawn.

Even now, I am startled that the image of a dead friend sounds
like a moonlit brook chewing its way down a desolate arroyo and smells
like corn tortillas in hot oil and frijoles sizzling in pork fat.

⁕

The Blue Nile flows from Bahir Dar to Khartoum, there it meets the
White, fallen from Victoria and Kampala. Together,
they twist through Thebes and Giza, then to Alexandria.

Here, the river spreads upon its delta and drowns
in the immense azure of the Mediterranean. Gone are waters Ugandan,
Ethiopian, Sudanese, Egyptian—only the flat blue arc

of the sea. Does the sea remember the mountains, the falls, the fast runs
through gorges thick with the flesh of death? Does the eagle recall the
deep red fear of the rabbit, or the infant lamb?

The waters of the Niles again make landfall near Marseilles,
where Parisians, grown thick on yellow sauces and purple wines, fatten
in the careless blue light of the European sun.

⁕

A fungus, turgid with old water, thick with spore, rises black
from the emptied chest of an old pitch pine. And mud like lizards
coils between broken stones, between mahogany and paintbrush,

into the fronds of fiddlehead ferns. Above, the wet breath
of the land gathers into virulent clouds, and driven
by plasmic winds, seeks the waters of its birth.

Beside the rivers below, priests in their dark habits gather
water with small blue bottles. And beneath a curiously carved archway,
a woman like old linen pulls salvation from a stone font,

and lays it carefully upon head, chest, and shoulders—the damp trace
of Golgotha, bleached of its path, slips, where she kneels,
into creases of forehead and trickles between her emptied breasts.

⁕

Her life was measured out in Mason jars, placed in precise rows
in plywood closets. Green glass filled with the bits

and buttons of her life. Clouded with age.

Tarnished brass rings, bubbled centers rimmed with broken
red rubber seals. A jar for each thing she'd done without
beneath the soiled canvas and the sudden rains, the yellow dust

and the blank face of Kansas. At age fifteen, she was wed
to a thirty-year-old man. She bore him eleven children,
then packed her life away in jars. One still festers

in memory, filled with sea water and sand and periwinkle shells,
one glimpse of a mythical eastern sea, sealed in a bottle and held
forever in a closet. Drained into a rusted iron sink when she died.

✳

Looping their way onto moonlit beaches, silvered grunion grope among
rippled sands for a spot to leave their jellied eggs and squeeze their
silken milt, for a place to stand and force a final dance beneath the
moon's

glistening indifference. Were it within my power to reverse the currents
that carried me to this awefull place, I would do it! So certain am I
that another's compass would have steered me to a deeper, darker,

holier space. A place where evil did not dress in black and dance
upon the tithes of the unborn, where justice was not wrapped
in darkness, a place where children play among snowy egrets, and

wolves feed freely upon the flesh of men. The grunion,
gilt with the fall of star's light, wriggle in the sand oblivious to all
but the motions of the tides and the rhythms of their lust.

✳

A simple pond in the deep of a nameless wood, brittle with the cold.
Ripples of the last dread wind, or the next, roll across its face. Still
now with the silence of the storm, the wolfen snowfall.

The light from a single star, Castor or Pollux, illumines
a single snowflake at the crest of a frozen wave, and that light shatters
into certain blues and elegant whites that float

among the packets of darkness like curious moths.
The light of a thousand other dying suns glistens in the frosted tops

of the ageless bristlecone pines, and the moon is eating the eastern sky.

The quietness is most complete now, the storm has stilled the winds
that curled behind my eyes, and the iced blade of this winter night
has finally slain the fey god that followed me here.

The river was cut by the world's great flood and runs over rocks from the basement of time. On some of the rocks are timeless raindrops. Under the rocks are the words, and some of the words are theirs.

I am haunted by waters.

— Norman Maclean